The Knights Templar Hiding in Plain Sight

The Life of Robert II Furfan de Ros

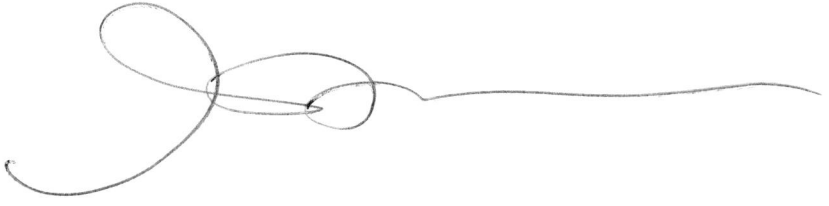

S. G. Rose

Copyright © 2020 S.G. Rose All rights reserved

No part of this book may be reproduced, or stored in a retrieval system, or transmitted in any form or by any means, electronic, mechanical, photocopying, recording, or otherwise, without express written permission of the publisher.

ISBN- 9798640561340

Cover design by: S. G. Rose

Editing: Mark David Rose

Library of Congress Control Number: Pending

Printed in the United States of America

Dedication

I would like to thank my children
Mark, Pierce, and Alyssa for
understanding my need to
work on this book for
10 years of their lives.
Yes, Robert II de Ros is my addiction
there is no other excuse.
Thanks to Zak, Seamus, Andrew
for being ready critics of my research and
answering my question
"Is this interesting to you?"
and saying yes.
To Craig Stevens for always being a ready ear.
To Kevin Conquest for allowing me
to tell him his family' story, as I understand it.
Thank You to Mick Harden MBE
for encouraging me and escorting me
around all the de Ros sites in
Yorkshire
and Understanding.

There are no such things as Coincidences, they are Realities you were not aware of.

Susan Rose

Contents

Forward	pages 13-17
#1 What's in a name?	pages 19-20
#2 One of the Knights Templar Effigies, Temple Church London	pages 21-22
#3 Could this have been a Cover Up?	pages 23-28
#4 My Conjecture	pages 29-30
#5 Who was George Rose?	pages 31-33
#6 Sir Walter Scott	pages 35-40
#7 Pagens, Paganels, de Payens, FitzPayne	pages 41-43
#8 Battle of Hastings 1066	pages 45-47
#9 Going to Scarborough Faire	pages 49-52
#10 For All the Money	pages 53-59
#11 Trois Bouts d'Leau	pages 61-64
#12 A Catherine's Wheel	pages 65-67
#13 Great Scrolls of the Board of Normandy	pages 69-70
#14 Married to the King of Scotland's Daughter	pages 71-72
#15 Children of King William the Lion	pages 73-74
#16 Hubert de Burgh	pages 75-79

#17 3rd Crusade with Richard the Lionhearted	pages 81-84
#18 Yorkshire Contingent	pages 85-86
#19 The Sheriff of Nottingham	pages 87-89
#20 So, He sold a bit of English Countryside to France	page 91-92
#21 Robert de Ros and his Abettors	pages 93-103
#22 A Magna Carta Surety	pages 105-106
#23 The Close Family Interconnections of the Surety Barons of the Magna Carta	pages 107-113
#24 Women's Rights in 1215	pages 115-116
#25 His Castle in the Poem Beowulf	pages 117-119
#26 America's First President/ Robert II de Ros	pages 112-123
#27 Problems/ 3rd Crusade	pages 125-127
#28 Patron Saints of the LGBTQ	pages 129-132
#29 Peacocks and Knights	pages 133-135
#30 Rules of the Knights Templar	pages 137-158
#31 Text of the Magna Carta	pages 159-171
Bibliography	pages 173-176

Forward

To me, the quest for knowledge of human history is the most important quest of all. The thirst for knowledge, is what makes us human. It changes us from being mere animals. The why of everything is empowering to us. So then if we have paid attention to History we will have learned from our predecessors, how to overcome our future problems. As they will not be very much different to what our ancestors have already overcome. I wanted to understand the foundations of Clan Rose. So, with a great enthusiasm I sat down with A Genealogical Deduction of the Family of Rose of Kilravock. A purely innocent start to a vastly interesting and amazing story. My early understanding was that Clan Rose was a small Clan in comparison to the Gordons, Frasers and MacDonalds. Not realizing what direction that my quest for the foundations of Clan Rose would take me in. I started investigating. I became surprised at every new-found source of power that time had forgotten about what the Rose family had done.

I did not suspect all the subsequent different directions it would lead me in. I found that the most important occurrences for Clan Rose would transpire well before to the year 1280 AD.

Prior to 1280 AD the Rose Barony had not been attached to the lands of Kilravock, but on a neighboring piece of land. The Rose Barony was not given for an Act of Bravery or nor seemingly anything of that sort, the de Ros/Rose Barony was attached to the land next to the lands of Kilravock, sometime prior to the year 1280.

I have now worked for years on uncovering the most amazing sets of documents. The Story of "The Knights Templar Hiding in Plain Sight," began to unfold much like an onion. Story by amazing story, making it very wondrous onion indeed.

April 15, 1746 Bonnie Prince Charlie was camped outside Clan Rose's Kilravock Castle awaiting to meet the Duke of Cumberland in an epic battle, between Prince Charlie's Jacobite Army and the Duke of Cumberland's Royalist Army representing the English Hanover King George. This battle was to take place on Culloden Moor just 6 miles away from Kilravock the very next day. This battle would decide which Royal family should be the rightful Kings of Scotland and England. Prince Charles Edward Louis John Stuart or George Louis of Hanover, King of England, and subsequently who would then still be governing Scotland and England to this day.

Prince Charlie, the day before the battle, while walking out with the Baron Hugh Rose and his wife to visit the orchard that Hugh was just putting in, states in a now famous quote; "How happy, Sir, you must feel, to be thus peaceably employed in adorning your mansion, whilst all the country round is in such commotion." Prince Charlie ends up staying for quite some time that day enjoying the company of Chief Hugh and his wife, long into the night.

The next day Prince Charles is soundly beaten during the disastrous Battle of Culloden. In the aftermath of that now infamous Battle, the Duke of Cumberland rides the 6 miles to Kilravock Castle on horseback, demanding an answer from Chief Hugh Rose, quote; "So you had my cousin Charles here yesterday?" Hugh Rose the 13th of Kilravock replied, that he could not prevent the visit. "Oh!" says the Duke, without alighting his horse. "You did perfectly right." General, Duke of Cumberland turned and rode off.

I had no knowledge yet as to who Hugh Rose's ancestors were. Just that they had to be someone important. All other Scottish Clan Castles belonging to Prince Charles's supporters were demolished, along with their documents which were subsequently confiscated and destroyed. Bonnie Prince Charlie just stops by and hangs out with the Roses the day before a major battle?

My next question was, if the Roses were not nobodies, then who exactly were they? My next wonderings were the Duke of Cumberland charges up on his horse and does not just slaughter Hugh Rose the

13th where he stood. During the chaos from the aftermath of the disastrous Battle of Culloden on April 16, 1746, tempers of the English Army reportedly were extremely high. The Duke of Cumberland had just given his army the order, "Put All to Fire & Sword!" upon the order of King George. "Put All to Fire & Sword!" Exactly what does that sentence mean?"

First said by English King Henry the 8th, in the 1500's about what Scotland needed in his day. The term is also referred to as "A Rough Wooing" or Rape.

King Henry the 8th's orders were simple, and terrifying - 'Put All to Fire and Sword'.

"...Put all to fyre and swoorde, burne Edinborough towne, so rased and defaced when you have sacked and gotten what ye can of it, as there may remayn forever a perpetual memory of the vengeaunce of God lightened upon them for their faulsehode and disloyailtye."

"Do what ye can out of hande...to beate down and overthrowe the castle, sack Holyrood house, and as many townes and villages about Edinborough as ye may conveniently, sack Lythe [Leith] and burn and subvert it and all the rest, putting man, woman and child to fyre and sworde without exception, where any resistance shallbe made agaynst you..."

So, "putting man, woman and child to fyre and sworde without exception," then means, put every man, woman and child. No exceptions. Except for the Roses. With the wholesale destruction of castles in Scotland also then destroyed most all the records and documents of Scottish History at that time.

By destroying the records and documents of Scotland, General Cumberland destroyed the History of Scotland, as if the Scottish Rebels never existed at all. Death and Destruction rained down on Scotland, surrounding Kilravock Castle, but never ever touching it. Wondering about the complete devastation of Scotland but not to the Rose Family, Why? My guess there had to be a set of very important Roses that I needed to locate.

Here is a document currently located in the Castle Belvoir Archives (owned to this day by the descendants of the de Ros family). The letter is from the Marquess of Granby to his father the Duke of Rutland in 1746. The Battle of Culloden occurred April 16, 1746. The Duke that Granby is referring to in this letter is the Duke of Cumberland.

Please note that The Marquess of Granby (England) is a relative of Baron Hugh Rose (Scotland.)

Marquess of Granby (on horse) by Edward Penny 1746 Public Domain

"The MARQUESS OF GRANBY to his father, the DUKE OF RUTLAND. [1746,] June J7. Fort Augustus. — "We have been at this place about a month, and I believe we shall stay here some time longer. The Duke since he has been here has sent out several detachments to drive in all the cattle belonging to the rebells, and to destroy and burn all their country, which they have performed with great success, having drove in several hundred head of cattle, and burnt everything they came near, without the least opposition. The Duke sent a detachment of an hundred of Kingston's horse, fifty on horseback and fifty on foot, into Glenmorrison's country to burn and drive in cattle, which they executed with great expedition, returning in a couple of days with a thousand head of cattle, after having burnt every house they could find. The Duke has now shown the gentlemen of Scotland who gave out that the Highlands were inaccessible to any but their own people, that not only the infantry can follow rebel Highlanders into their mountains, but that horse upon an occasion commanded by him find nothing impracticable. Captain Chadwicke who commanded says he was surprised to see the boldness of the men who galloped up and down mountains that he thought was impossible to have walked down. I hear some of our new raised regiments are continued; if mine should be continued I'll get to Newcastle as soon as possible, for Stanwix writes me word that our men begin to be uneasy and want their discharge." H. Fox to the DUKE OF RUTLAND. 1746, June 17. — Concerning Lord Granby's regiment. The SAME to the SAME. 1746, June 21. Holland House. — I am glad to find by your letter of the 18th that the (Marquess of Granby's) regiment is in good temper. (Public Domain)

What's in a name?

Chapter 1

Reading "A Genealogical Deduction of the Family of Rose of Kilravock: With Illustrative Documents From the Family Papers, and Notes," by Hugh Rose starts in year 1280. Filled with amazing people. But with not enough power to stop the English Government Forces from laying waste to them. There had to be more.

Kilravock Castle Scotland, Photographed by author in 2019

Acquiring knowledge about people who lived in the 12th and 13th Centuries is difficult at best. Most people rely of other people's work, which can lead to absolute falsehoods being reported on what someone did or did not do over 800 years ago. After uncovering the first clue I sought only original source documents as my primary references for research. For most of my document searches I sourced Church registers. Documents in churches are the listing of the Churches congregations not a political document. Meaning these documents were written when the actual occurrence transpired.

I first concentrated on researching the name of the family Rose which led to a complete dead end. Nothing showed up prior to year 1280AD. So, I looked closer at the Rose name spelling in the old Clan Rose history book. I find that "de Ros" was the answer…the last name of Hugh Rose the First Baron, was sometimes being listed as de Ros and not a misspell, but Rose was a disguise, Hiding him in plain sight. The Documents were not edited or corrected by scribes prior to having a Chief's Seal being placed upon it. There appears to be no rough drafts like we have today. During this time in history, there seemingly were no qualifications for being a scribe as a job and no one was educated much outside of the Clergy. They did not know that the words were "not" being transcribed in a clear and thoughtful manner. Sometimes the documents even contain literal scribbles interspersed with real misspelled words, but scribbles none the less. Were the scribes being paid by the word or scribble? Since the Chiefs were not educated in reading their own language, how would they know if the scribe was any good?

So, with that in mind, I went back to researching the last name of de Ros. That name would be the start of my research going backward in time. It seems the de Ros family had originated in Normandy, France. As in, they participated in the Norman Invasion of England and fought alongside those famous Norman Knights fighting for William the Conqueror during the Battle of Hastings, in year 1066 AD. So, then using "de Ros" as my new avenue to research I find this: "Robert II "Furfan" deRos, one of the 25 Surety Barons of the Magna Carta, and then Robert II "Furfan" de Ros one of the Famed "9" Knights Templar Effigies located in Temple Church on Fleet Street, London." Plus, he had the exact same Heraldry as Clan Rose in Scotland. I could have stopped my research right there…but no that would not be like me, I seem to be unable to let well enough alone.

One of the Knights Templar Effigies, Temple Church London

Chapter 2

"His effigy is described by Gough, in "Sepulchral Monuments," as the most elegant of all the figures in the Temple Church, representing a comely young knight in mail, and a flowing mantle with a kind of cowl; his hair neatly curled at the sides; his crown appears shaved. His hands are elevated in a praying posture, and on his left arm is a short, pointed shield charged with three water-bougets. He has on his left side a long sword, and the armor of his legs, which are crossed, has a ridge, or a seam up the front, continued over the knee. At his feet is a lion, and the whole figure measures six feet two inches..."

The Effigies in Temple Church London are carved from stone. They are portrayed in their 30's ready to bounce back to life and defend Jerusalem when Christ returns. They are not memorials of lives past but visions of their life yet to come. To be ready to respond to the returned Christ's request for the Army of Christ.

As to portraying Robert II as a Knights Templar, it is not. It is a portrait of Robert II de Ros yet to be. I have read critiques about him not being a Knights Templar based on the carving in Temple Church, then of course he is not correctly portrayed, as that was not the intent of the artist commissioned to carve the Effigy.

The de Ros Effigy is described as the best effigy in Temple Church, with the most detail and best use of marble of all the effigies. Why would he be the best? After my research in the de Ros/Rose source of power and the family connections to the de Payen family, specifically Hugh de Payen. It would make sense as to why the de Ros effigy would be the Best Effigy in Temple Church.

Robert II "Furfan" de Ros Effigy Temple Church England.
photographed by author Susan Rose 2017

Could this have been a Cover Up?

Chapter 3

The History of the Knights Templar By Charles G. Addison, [1842]
(Public Domain)

"By the side of the earl of Pembroke, towards the northern windows of the Round of the Temple Church, reposes a youthful warrior, clothed in armour of chain mail; he has a long buckler on his left arm, and his hands are pressed together in supplication upon his breast. This is the monumental effigy of ROBERT LORD DE ROS and is the most elegant and interesting in appearance of all the cross-legged figures in the Temple Church. The head is uncovered, and the countenance, which is youthful, has a remarkably pleasing expression, and is graced with long and flowing locks of curling hair. On the left side of the figure is a ponderous sword, and the armour of the legs has a ridge or seam up the front, which is continued over the knee, and forms a kind of garter below the knee. The feet are trampling on a lion, and the legs are crossed in token that the warrior was one of those military enthusiasts who so strangely mingled religion and romance, **"whose exploits form the connecting link between fact and fiction, between history and the fairy tale."**

"In an ancient genealogical account of the Ros family, written at the commencement of the reign of Henry the Eighth, A.D. 1513, two centuries after the abolition of the order of the Temple, it is stated that Robert Lord de Ros became a Templar, and was buried at London. The writer must have been **mistakened**, as that nobleman remained in possession of his estates up to the day of his death, and his eldest son, after his decease, had livery of his lands, and paid his fine to the king in the usual way, which would not have been the case if the Lord de Ros had entered into the order of the Temple. He was doubtless an associate or honorary member of the fraternity, and the circumstance of his being buried in the Temple Church probably gave rise to the mistake."

"Robert Lord de Ros, in consequence of the death of his father in the prime of life, succeeded to his estates at the early age of thirteen. In the second year of the reign of Richard Cœur de Lion, (A.D. 1190,) he paid a fine of one thousand marks, (£666, 13s. 4d.,) to the king for livery of his lands. In the eighth year of the same king, he was charged with the custody of Hugh de Chaumont, an illustrious French prisoner of war, and was commanded to keep him safe as his own life. He, however, devolved the duty upon his servant, William de Spiney, who, being bribed, suffered the Frenchman to escape from the Castle of Bonneville, in consequence whereof the Lord de Ros was compelled by King Richard to pay eight hundred pounds, the ransom of the prisoner, and William de Spiney was executed."

"On the accession of King John to the throne, the Lord de Ros was in high favour at court." "He was sent into Scotland with letters of safe conduct to the king of Scots, to enable that monarch to proceed to England to do homage, and during his stay in Scotland he fell in love with Isabella, the beautiful daughter of the Scottish king, and demanded and obtained her hand in marriage."

"From his sovereign the Lord de Ros in the year 1213 he was made sheriff of Cumberland. He was at first faithful to King John, but, in common with the best and bravest of the nobles of the land, he afterwards shook off his allegiance, raised the standard of rebellion, and was amongst the foremost of those bold patriots who obtained MAGNA CHARTA."

"He was chosen one of the twenty-five conservators of the public liberties, and engaged to compel John to observe the great charter."

"Upon the death of that monarch he was induced to adhere to the infant prince Henry, through the influence and persuasions of the earl of Pembroke, the Protector."

"He died in the eleventh year of the reign of the young king Henry the Third, (A.D. 1227,) and was buried in the Temple Church."

"The above Lord de Ros was a great benefactor to the Templars. He granted them the manor of Ribstane, and the advowson of the church; the ville of Walesford, and all his windmills at that place; the ville of Hulsyngore, with the wood and windmill there; also all his land at Cattail, and various tenements in Conyngstreate, York. The principal benefactors to the Templars amongst the nobility were William Marshall, earl of Pembroke, and his sons William and Gilbert; Robert, lord de Ros; the earl of Hereford; William, earl of Devon; the king of Scotland; William, archbishop of York; Philip Harcourt, dean of Lincoln; the earl of Cornwall; Philip, bishop of Bayeux; Simon de Senlis, earl of Northampton; Leticia and William, count and countess of Ferrara; Margaret, countess of Warwick; Simon de Montfort, earl of Leicester; Robert de Harecourt, lord of Rosewarden; William de Vernon, earl of Devon."

"King John was resident at the Temple when he was compelled by the barons of England to sign MAGNA CHARTA. Matthew Paris tells us that the barons came to him, whilst he was residing in the New Temple at London, "in a very resolute manner, clothed in their military dresses, and demanded the liberties and laws of king Edward, with others for themselves, the kingdom, and the church of England."

You will notice there are a few discrepancies between my research into Robert II de Ros and in this above-mentioned book by Charles Addison, dated 1842. Mr. Addison leaves out quite a few important details and seemingly does not investigate more important ones. No mention of Robert being with King Richard on the 3rd Crusade. No mention that King Richard wore the sur coat of the Knights Templar on the 3rd Crusade. That Robert was completely in the Knights Templar service during that same Crusade as a Fighting Knight and then became an Associate Member upon his return to England.

Knights Templar fighting knights were not first-born sons as Robert was. Robert had more influence being an associate member than as a fighting member. I have included the Rules of being a Knights Templar at the back of this book. Mr. Addison appears to be trying to discredit Robert out of hand, but why? The Knights Templar at Temple Church London at the time of King John were also members of the First Barons Revolt and then became some of the 25 Sureties of the Magna Charta. Temple Church was not a Haven of Safety for King John but a site of House Arrest.

When Mr. Addison first wrote this book on the Knights Templar, he obviously did not have easy internet access to important documents of the day but, none the less he had access to them if he wanted. The sources that he used are not original source documents at all, but a rehash of mis-quotes from multiple sources close to the time that he wrote his own book. He must have researched the history of such an important man as Robert II de Ros prior to writing his book "The History of the Knights Templar, dated 1846." In it he quotes Matthew Paris, but Matthew Paris was only 15 years old at the time of the Magna Carta? Matthew Paris has since been discredited on many of his writings as being not factual by modern day historians.

My questioning of Mr. Addison's work is this. As an English Barrister, Addison had access to the documents contained at Temple Church. Why then did he seemingly never access them? Or had they gone missing in 1772, when George Rose was organizing the Scroll Room?

When one has an inkling as to the powerful family connections of Robert, one then would not question as to why Robert is in Effigy in Temple Church. Yet he was in Plain Sight laying right there next to Pembroke in beautiful life size Effigy. Were the de Ros documents covered up on purpose? Was Mr. Addison asked to not write about what he knew? It seems strange to me that a complete blank is thrown over the top of Robert, by Mr. Addison.

Why would that be? The only thing I can reason out is that the Barons and Robert seemingly ran "rough shod" over all the powers that be, back in the day. Much like what a Catherine's Wheel did: "As one that runs rampant." But still why would Addison leave these important items out or boldly change others?

Were the Scrolls pertaining of the activities of the de Ros family an obsession for George Rose and he then removed the documents/scrolls in 1772? George had full unobstructed access to all the documents. Was George Rose wanting to hide the activities of Robert during the time of the Barons Revolt and Magna Carta? Or was he just wanting to have in his possession the scrolls documenting the activities of his own predecessors?

What I do know is that Sir Walter Scott admits to having some of these Documents in Scotland. Exactly how they were moved from London to Scotland, would be a guess now. Who gave them to him? George, or his son William Rose? I can only offer conjecture. Both Roses had access to the archives and at some point, they ended up in the hands of Sir Walter Scott. The act of lying and subversion was not invented in the last century, and it is perfectly reasonable that 1700's archivists would be just as prone to omission to cover their actions as we are today. Currently, documents pertaining to the Troubles in Ireland that were supposedly in the Archives in the 1990's yet they too have also gone missing, meaning it is common for classified documents to disappear.

Missing from Addison's book are Robert's important close family connections to the Knights Templar's first Grand Master Hugh de Payn's his own family. When it is a well-known fact that the original 9 Knights Templar were related by Birth or Marriage. He also incorrectly lists Robert meeting and subsequently marring King William the Lion's daughter as occurring in the year, 1200AD. Robert had been married to his wife Isabella for 10 years prior to his escorting King William into London in the year 1200 to meet with the English King. Then subsequently escorting his Father-in-Law to Bonneville sur Touques Castle, Robert's own Castle in Normandy.

He also lists incorrectly the amount of Roberts fine for allowing Hugh de Chaumont to escape. When the fine is well documented to this day in London. Rievaulx Abbey being the first gift to the Knights Templar from Robert is never mentioned at all, again why? These especially important items to the Knights Templar Organization could have only been left out on purpose.

At the end of this book I have for you to read Rules for the Knights Templar, please reference Article #69 of the Laws for Knights Templar and how it would pertain to Robert. In that Robert, can work within the Knights Templar in an advisory capacity. Robert who as a "First Born" son, was more useful to the Knights Templar Organization as a landed Baron of England & Normandy, than being an ordinary fighting Knight of the Temple order. In as such Robert would have exerted pressure on the Prince and future King John, King Richard, King Henry III and King William the Lion of Scotland plus the two Popes, Innocent III & Honorius III. To which history has shown as to have absolutely taken place.

The Knights Templar organization needed "Inside Men," to assist in getting contracts and or negotiating for the Knights Templar, with these 4 important Kings and 2 Popes, Robert seems to have worked for the betterment of the Knights Templar Order. And yet this is not mentioned to this day, why?

So then, what is the reason for Mr. Addison discrediting to Robert? Was Mr. Addison just lazy in his research? The outcome of what Robert did in his lifetime, did affect England, the Knights Templar organization itself and the World at large. But Mr. Addison did not want that to be common knowledge. It is a complete disregard to History.

My Conjecture

Chapter 4

I must state this Chapter is my conjecture. My questions arise as Charles G. Addison describes the Effigy of Robert II de Ros located in Temple Church as a "mistake." "Whose exploits form the connecting link between fact & fiction." In 1842 and 1852 Addison wrote about the Knights Templar and Temple Church. But a George Rose had prior access to the Westminster Scroll room and the Law Library located in Temple Church to re-organize them, in 1772. Since the forward to the second volume of the scroll documents contains a "sincere thank you to George Rose for his diligent work on the Scrolls and Documents, **"too bad there are so many scrolls missing."**

Where did the missing scrolls go? A particularly good friend of George Rose's son William was Sir Walter Scott who wrote and published Ivanhoe. Then later recanted and admitted he was working off old Manuscripts located in his Oak Cabinet, when called out for changing the pertinent names of people and place names from history.

A huge controversy played out in newsprint over this very topic as Historians of the era noticed the similarities of historic fact in the book Ivanhoe dated 1818. Scott changed site names such as Sherwood Forrest for Englewood Forrest referenced in the Robin Hood tales dating from the 1200's. Stories that had been told orally for 200 years previously to them being written down in the 1400's. Walter Scott using the title of "Sheriff of Nottingham" then changing the meaning of that title. "Of Nottingham" was really where the sheriff was from, not where he was policing.

Then using King Richard, the Lionhearted in the Robin Hood

storyline. King Richard and Robert II de Ros, yes, did know each other. Yes, Robert II de Ros was the Sheriff (de/from) Nottingham in Englewood Forrest in Cumberland 1214, where Robin Hood was first working with Little John, as told in the first oral renditions of Robin Hood. King Richard was never in Cumberland or Nottingham.

The scrolls pertaining to Robert II de Ros are not to be found in Westminster or the Law Library at Temple Church. Were they purposely removed by a George Rose in 1772? Then possibly turned over to Sir Walter Scott by George Rose's son, William Stewart Rose? William Rose was condemned by his constituents at the time, as acquiring his jobs due to the nepotism of his father, George Rose. William's jobs were considered "Sinecure." Government secured pensioned jobs that William did zero work for. William is listed as a poet and friend of the poet Byron to whom he wrote poems, based on the stories about Don Juan (act of plagiarism). They were listed as William's own poems but, based on another poet's work.

What does Sinecure mean? Latin = Sine Cura = Without Care.

Robert II Furfan de Ros did lead an amazing life. I accessed Documents and Scrolls from Normandy, Papal Documents, Churches, Abbeys, and Manor Houses where the de Ros families held properties; I have located incredibly informative original source documents pertaining to the life of Robert.

Who was George Rose?

Chapter 5

George Rose, Public Domain

He achieved quite an amazing career in Public Service for being a son of a poor Minister.

He joined the Royal Navy in 1762, he was furloughed after being injured in battle.

In 1772 he was appointed "Keeper of the Records" Westminster as an assistant, then as Keeper himself after the retirement of his boss.

In 1776 he was made Secretary of the Board of Taxes.

In 1782 was designated Secretary of the Treasury

In1784 he is appointed by Duke of Northumberland MP of Launceston Cornwall.

In 1788 appointed Clerk of the Parliaments In 1790 MP of Lymington & Christchurch

In 1804 Joint Paymaster of the Forces & Vice President of the Board of Trade.

In 1807 The Duke of Portland appoints him Treasurer of the Navy & Vice President of the Board of Trade.

During his 1772 job Assisting the Keeper of the Records of Westminster, then becoming the "Keeper" Himself with no oversight during that tenure. Upon the publication of the Contents of the Scroll rooms in book format. The forward in that book commended George Rose for doing a fine job of the organization of all the Documents and Scrolls but, **"Too bad so many Documents and Scrolls went Missing."**

George Rose was a close friend and Attorney for of Admiral Lord Nelson. Nelson asked George Rose to board HMS Victory prior to the ship sailing for the Battle of Trafalgar; what Nelson wanted with Rose was a confirmation that should he be killed his wife Lady Hamilton and their daughter Horatia would be looked after by Rose. Rose was thus the last man in England to see Nelson alive. After Nelson's death Rose became Emma Hamilton's executor and Horatia's guardian.

George Rose's elder son, Sir George Henry Rose (1771–1855), was in

Parliament from 1794 to 1813, and again from 1818 to 1844, and he was the British Minister at Munich, at Berlin, and at Washington. He was made a Knight Grand Cross of the Royal Guelphic Order and in 1818 succeeded his father as Clerk of the Parliaments. He was the father of Field Marshal Hugh Rose, Baron of Strathnairn, who was described as one of the bravest men in the British Army and the best commander in the Indian Mutiny.

Georges second son the poet William Stewart Rose was great friends with Sir Walter Scott. William Rose was successively appointed:

1997 - 1800 Surveyor of Green-Wax Monies,

1797 - 1837 Clerk of the Pleas, "Sinecure,"

1800 - 1824 Reading Clerk to the House of Lords

1796-1800 partnering his father as MP for Christchurch

His post as Clerk of Pleas was considered to be a "sinecure" from his father's positions. Sinecure Meaning: Latin = Without Care A position requiring little or no work but giving holder status or financial benefit.

All three of these Rose men all had access to the Documents and Scrolls of early English History.

Sir Walter Scott

Chapter 6

Sir Walter Scott, Public Domain

The next few paragraphs are from the Forward to the Book Ivanhoe written by Sir Walter Scott. In these paragraphs he is explaining why he used original source documents and scrolls to pull his story of Ivanhoe in a tongue in cheek style. The Scottish FSA Antiquarians had been very vocally pointing out that he was changing real historic events and people in his Fictitious story of Ivanhoe, by changing the names and locations of real events. Introduction to the Book Ivanhoe:

"If the author, who finds himself limited to a particular class of subjects, endeavours to sustain his reputation by striving to add a novelty of attraction to themes of the same character which have been formerly successful under his management, there are manifest reasons why, after a certain point, he is likely to fail. If the mine be not wrought out, the strength and capacity of the miner become necessarily exhausted. If he closely imitates the narratives which he has before rendered successful, he is doomed to "wonder that they please no more." If he struggles to take a different view of the same class of subjects, he speedily discovers that what is obvious, graceful, and natural, has been exhausted; and, in order to obtain the indispensable charm of novelty, he is forced upon caricature, and, to avoid being trite, must become extravagant."

"It is not, perhaps, necessary to enumerate so many reasons why the author of the Scottish Novels, as they were then exclusively termed, should be desirous to make an experiment on a subject purely English. It was his purpose, at the same time, to have rendered the experiment as complete as possible, by bringing the intended work before the public as the effort of a new candidate for their favour, in order that no degree of prejudice, whether favourable or the reverse, might attach to it, as a new production of the Author of Waverley; but this intention was afterwards departed from, for reasons to be hereafter mentioned."

"The period of the narrative adopted was the reign of Richard I., not only as abounding with characters whose very names were sure to attract general attention, but as affording a striking contrast betwixt the Saxons, by whom the soil was cultivated, and the Normans, who still reigned in it as conquerors, reluctant to mix with the vanquished, or acknowledge themselves of the same stock. The idea of this contrast was taken from the ingenious and unfortunate Logan's tragedy of Runnamede, in which, about the same period of history, the author had seen the Saxon and Norman barons opposed to each other on different sides of the stage. He does not recollect that there was any attempt to contrast the two races in their habits and sentiments; and indeed it was obvious, that history was violated by introducing the Saxons still existing as a high-minded and martial race of nobles."

"After a considerable part of the work had been finished and printed, the Publishers, who pretended to discern in it a germ of popularity, remonstrated strenuously against its appearing as an absolutely anonymous production, and contended that it should have the advantage of being announced as by the Author of Waverley. The author did not make any obstinate opposition, for he began to be of opinion with Dr Wheeler, in Miss Edgeworth's excellent tale of "Maneuvering," that "Trick upon Trick" might be too much for the patience of an indulgent public, and might be reasonably considered as trifling with their favour."

"The book, therefore, appeared as an avowed continuation of the Waverley Novels; and it would be ungrateful not to acknowledge, that it met with the same favourable reception as its predecessors." "Such annotations as may be useful to assist the reader in comprehending the characters of the Jew, the Templar, the Captain of the mercenaries, or Free Companions, as they were called, and others proper to the period, are added, but with a sparing hand, since sufficient information on these subjects is to be found in general history."

"Ivanhoe was highly successful upon its appearance, and may be said to have procured for its author the freedom of the Rules, since he has ever since been permitted to exercise his powers of fictitious composition in England, as well as Scotland."

"DEDICATORY EPISTLE TO THE REV. DR DRYASDUST, F.A.S. Residing in the Castle-Gate, York."

"Much esteemed and dear Sir, It is scarcely necessary to mention the various and concurring reasons which induce me to place your name at the head of the following work. Yet the chief of these reasons may perhaps be refuted by the imperfections of the performance. Could I have hoped to render it worthy of your patronage, the public would at once have seen the propriety of inscribing a work designed to illustrate the domestic antiquities of England, and particularly of our Saxon forefathers, to the learned author of the Essays upon the Horn of King Ulphus, and on the Lands bestowed by him upon the patrimony of St Peter. I am conscious, however, that the slight, unsatisfactory, and trivial manner, in which the result of my antiquarian researches has been recorded in the following pages, takes the work from under that class which bears the proud motto, "Detur digniori". On the contrary, I fear I shall incur the censure of presumption in placing the venerable name of Dr Jonas Dryasdust at the head of a publication, which the more grave antiquary will perhaps class with the idle novels and romances of the day. I am anxious to vindicate myself from such a charge; for although I might trust to your friendship for an apology in your eyes, yet I would not willingly stand conviction in those of the public of so grave a crime, as my fears lead me to anticipate my being charged with."

"Of my materials I have but little to say. They may be chiefly found in the singular Anglo-Norman MS., which Sir Arthur Wardour preserves with such jealous care in the third drawer of his oaken cabinet, scarcely allowing any one to touch it, and being himself not able to read one syllable of its contents. I should never have got his consent, on my visit to Scotland, to read in those precious pages for so many hours, had I not promised to designate it by some emphatic mode of printing, as {The Wardour Manuscript}; giving it, thereby, an individuality as important as the Bannatyne MS., the Auchinleck MS., and any other monument of the patience of a Gothic scrivener. I have sent, for your private consideration, a list of the contents of this curious piece, which

I shall perhaps subjoin, with your approbation, to the third volume of my Tale, in case the printer's devil should continue impatient for copy, when the whole of my narrative has been imposed." "Adieu, my dear friend; I have said enough to explain, if not to vindicate, the attempt which I have made, and which, in spite of your doubts, and my own incapacity, I am still willing to believe has not been altogether made in vain. "The last news which I hear from Edinburgh is, that the gentleman who fills the situation of Secretary to the Society of Antiquaries of Scotland, is the best amateur draftsman in that kingdom, and that much is expected from his skill and zeal in delineating those specimens of national antiquity, which are either mouldering under the slow touch of time, or swept away by modern taste, with the same besom of destruction which John Knox used at the Reformation. Once more adieu; "vale tandem, non immemor mei". Believe me to be, Reverend, and very dear Sir,

Your most faithful humble Servant.

Laurence Templeton (aka Sir Walter Scott)

Toppingwold, near Egremont, Cumberland,

Nov. 17, 1817."

At the time that Ivanhoe was first released Sir Walter Scott published it without his name as the author but as an "absolutely anonymous production." The Antiquity Society of Scotland objected to sections of Ivanhoe as twisted historical facts. Yes, it looks to me that Scott is working from the scrolls in the Oak Cabinet. How could he have obtained the Scrolls? The son of George Rose named William Rose was a particularly good friend of Scott. As commented on in George Rose's work in the Scroll room in Westminster. "To bad so many scrolls went missing."

What Scott is then actually explaining was that yes, he used manuscripts/scrolls as a device to write and sell the book Ivanhoe. Why else would he use an "absolutely anonymous production," on the original printing of Ivanhoe. Subsequent editions were released with a much longer forward version explaining away his actions.

Pagen, Pagens, Paganels, de Payens, Payne, Paine or FitzPayne

Chapter 7

FitzPagen aka de Trusbut is the source of the Trois Bouts d'Leau heraldry used by the de Ros and Rose Families. The Great Grandfather of Robert II de Ros was the brother of Hugh de Payen the Original Grand Master of the Knights Templar... The connotations of this name bring in a whole new angle into the view of Robert II de Ros' life.

Note the term "Fitz", means son of...de Pagan or de Payens or de Payne, de Payens, the originator of the Knights Templar organization was named Hugh de Payens.

Research shows the dePayens name gradually changed from its original form of Pagani, Pagamis to Pagan, Pagen, Paven, Payen, Payne or Paine. Also, Payson and Pyson, Payson and others, all of which were merely different forms of the same appellation. There was no standardization of spelling, yet. In the Domesday Book compiled in 1086 AD, the name was uniformly written as Pagan.

There does seems to be some sort of awkward implications with that name...Pagan, it seems that back in the day, people were baptized in the local river, stream, or lake etc. Any source of water would be used in northern Europe for baptism. Diseases abounded at that time and the water could be a source of death or it could be literally freezing. People were referred to as "Pagan" for not subjecting themselves for the dunk in the local water. Not necessarily meaning they rejected Christianity, they were opposed to the local water.

King Baldwin II ceding Temple Mount to Hugh de Payns by Guillaume de Tyr 1300's Public Domain

All the 9 original Knights Templar men were related by birth or marriage... Of the original 9 Knights Templar were, Raymond IV of Saint-Gilles; Count of Toulouse; Bohemond; Duke of Taranto; Godfrey of Bouillon; Hugh dePayens, Count of Vermandois; and Robert Duke of Normandy aka Robert Cut-hose, William the Conqueror's eldest Son, commanded this small army. Bishop Adhemar of le Puy, the close friend of Pope Urban II, was their spiritual leader. The Duke of Normandy, Robert Curthose, French Robert Courteheuse (born c. 1054—died February 1134, Cardiff, Wales), duke of Normandy (1087–1106), a weak-willed and incompetent ruler whose poor record as an administrator of his domain was partly redeemed by his contribution to the First Crusade (1096–99). The eldest son of William I the Conqueror, Robert was recognized in boyhood as his father's successor in Normandy. Nevertheless, he twice rebelled against his father (1077/78 and c. 1082–83) and was in exile in Italy until he returned as

Duke on his father's death in 1087. He was totally unable to control his rebellious vassals or to establish a central authority in Normandy.

In 1091 Robert's younger brother, King William II of England, invaded Normandy, and compelled Robert to yield two counties. William attacked again in 1094, and when a peace was made that gave him control of Normandy in return for money, Robert joined the First Crusade. He fought at Dorylaeum (1097) and was at the capture of Jerusalem (1099). His courageous leadership contributed to the victory at Ascalon (1099). --Encyclopedia Brittanica

So, then my take is this, Robert Curthose was then very possibly related to the de Trusbut/de Payens as well. The implication then is Robert II de Ros has 3 family crosses into the de Payens family at least. Robert II, then could have been related in some manner to William the Conqueror, who was not a legitimate son of a marriage. "To keep it in the family, so to speak." This seems to be an ongoing theme throughout this book so far, Family Connections.

Any which way of it, every list of the 8 or 9 original Knights Templar states that they were related by blood or marriage. All through the Pagani, Payens, Payne etc. family line, no matter how the scribe wrote their names it was still only one family. Bernard of Clairvaux was also a member of that same great family. Bernard of Clairvaux was the Cardinal who was asked to write up the rules of the Knights Templar by the Pope. He also designed Robert II de Ros family's Rievaulx Abbey, coincidence? No, I would doubt that a mere coincidence had anything to do with this.

There are no such things as Coincidences only Realities you were not aware of.

Battle of Hastings 1066

Chapter 8

In the year 1066 AD, William the Conqueror invaded England with his Norman Knights and took over England. There are multiple accountings of a set of five "de Ros" knights, who were listed as part of William's invasion force that attacked and sacked England. William the Conqueror removed (a politically corrected term for annihilated) most all male English landed gentry and installed his own Norman Knights in their place and made them England's new Barons and Earls. Yes, I mean that literally. William kept the women but traded out their English husbands for his own Norman Knights. Brutal? Yes. Maybe that is why we still to this day; call him the Bastard, figuratively and literally speaking.

I researched these five "de Ros" Knights. The 5 de Ros's are listed on the famous Bayeux Tapestry. The Bayeux Tapestry is a long woven and embroidered tapestry showing the Norman Knights with their own heraldry shields in action during the Battle of Hastings in 1066 AD. These same five de Ros knights are also listed on every other recognized accounting list of Norman Knights who had accompanied William the Conqueror in 1066AD. The de Ros's all originated from the same small town called "Rots" in Normandy, France. The town of Rots is located on the peninsula that sticks out on the lower portion of Normandy Beach. On a map of Normandy Beach, it would look like a thumb sticking out.

My research expands into the French and German language meanings of names, as Normandy was considered either French or German changing back between them both after every conflict in History. Research concludes that, "de" means "of," so then "de Ros", would not be a "sur" name as such. I considered what the word "Ros" means as well. It means Red.

So, then these five de Ros guys listed on the Battle Abbey Roll, who had lived on the peninsula and were from the town of Rots, that was located on the promontory sticking out of Normandy Beach. Real sur names back in this time in history were largely not used, a person was identified by his given or first name plus a descriptor name, as to where they were from or what they did or what they owned. It has nothing to do with the flower Rose at all but their "RED" Hair or Complexion whichever.

I find "first" name problems with these five de Ros Knights, on all separate listings of the Norman Invasion Force with the listings of their first names. They seem not to have been recorded correctly, I find them to have been listed with different given names on most every Norman Invasion "Battle" list. But the actual descriptions of them are the same for all the 5 de Ros brothers. So now my research goes into why the discrepancy of the first names of these same "five" de Ros Norman Knights.

OK work with me here… so now just imagine if you will on Invasion Day Eve, 1066 AD your job, as one of the scribes for William "soon to be the Conqueror." You have the job of documenting his Invasion Force of his 5000 Fighting Knights. First problem as a Scribe is the non-standardized form of spelling, which will inhibit your work. Spelling standardization was not organized by any European Country at that time and would not to be for another 300 years. Taking all that into account now, you the Scribe are tasked with the job of getting the first name, the descriptor or sur name, plus a drawing of the Heraldry Shields for each one of the 3,000 Fighting Knights to be listed.

Just to complicate your day, you must navigate through the entire Invasion Force of 30,000 people. Knowing that if you mess up, possibly soon, a well-armed and incredibly angry Knight will be looking you up, because you possibly missed his name on the Official Invasion Force List. Just a little bit pressure on you, Scribe? You the Scribe are now hating life; you are going from surly knight to surly knight asking for their names and drawing their heraldry. After roughly 3,000 testosterone filled, well-armed knights screaming at you to get lost, a few of the knight's "Given Names" might have become slightly jumbled to say the least. Not your fault, you have a terrible job being a Scribe in 1066 AD. It's not your fault at all, the deck was stacked against you succeeding at all.

I am now only going to focus of just one of those five "de Ros" knights by the name of Guilliume de Ros/de Rots. His first-born son named Piers de Ros born in England, after the 1066 Invasion. Of course, I investigated all 5 de Ros brothers, only one of these 5 guys from Rots, in Normandy France, goes on to be the ancestor to a famous person that time forgot.

Going to Scarborough Faire

Chapter 9

This same Piers de Ros grows up to be the Castellan of Bonneville-sur-Toques in Normandy France, as a hereditary holding. Meaning that his father before him was also Castellan of this same Castle, and it is listed as belonging to William the Conqueror, as William's "Home".

Castellan Definition: Middle English castelleyn from Latin castellanus "occupant of a fortress"

Bonneville-sur-Toques Castle, is listed as being owned by William the Conqueror, located in Normandy, France. All Castles were owned by the current King, period. To occupy a castle, one must be a member of the King's army and pay taxes on the produce of the land. The Taxation Roll Documents that show the taxes paid by this same splinter of the de Ros family for generations are still to this day held by the British Government. Naming each subsequent first son to first son on the British Taxation Rolls. Bonneville-sur-Toques Castle continued under de Ros's family control through more than 7 generations to Robert II de Ros's first-born son William de Ros.

Bonneville-sur-Touques Castle had also been William Conqueror's own live-in Castle. Meaning William, the Conqueror put Piers' father, Guilliome de Rots, in charge of William's own personal Castle. Making the de Ros family extremely high up on the "Food Chain", so to speak of Norman Knights.

The first Castle of Bonneville-sur-Touques was built by Charlemagne to guard and be an overlook to the entrance of the Touques River to the sea. In the 1000's it was rebuilt as a Norman castle with five watch towers and a surrounding deep ditch with impenetrable walls for the

Duke of Normandy. This castle would become infamous in English History for hosting Kings. Bonneville-sur-Touques Castle would house King William the Conqueror, English King Henry I, King Henry II, King Richard the Lion Heart, King John, and Scottish King William the Lion.

William the Conqueror is responsible for more than 500 Castles being built in Normandy and Britain. He could not possibly manage all those Castles by himself. He rewarded his Norman Knights with these many Castles and the income from the field production from the Estates surrounding them, to give his best Fighting Knights working capital in exchange for their Knightly service when needed by the King. Remember that point…it is especially important. It is what Kings did, during this time in History. Piers de Ros was also the Steward or Castellan, the same definition applies but using an English word now, of Scarborough Castle. Scarborough Castle was then the Main Port of Entry in England for most all incoming goods and supplies from Europe. Meaning, Piers de Ros was controlling major amounts of currency and goods going through his port. A considerable honor bestowed on de Ros.

Ruins of Scarborough Castle, by author Susan Rose 2017

"Going to Scarborough Faire, Parsley Sage,

Rosemary and Thyme…"

You know the song come on, sing along with me. That song is a compilation of poems from the Scarborough area. The Faire was held after products arrived from Normandy. Yes, I looked that up.

Piers de Ros married Adeline d'Espec, sister and heir of Walter d'Espec, Lord of Hamlake. His marriage with Adeline d'Espec founded the vast "Land Fortunes" of the de Ros family. Adeline brought with her in marriage the Lordships of Hamlake, York, Warke, and Northumberland. The first son of Piers de Ros was named Robert I de Ros, Lord of Hamlake. Robert I was married to Sybil de Valoines. Robert I de Ros is listed as The Munificent Benefactor of the original Knights Templar organization.

For All the Money

Chapter 10

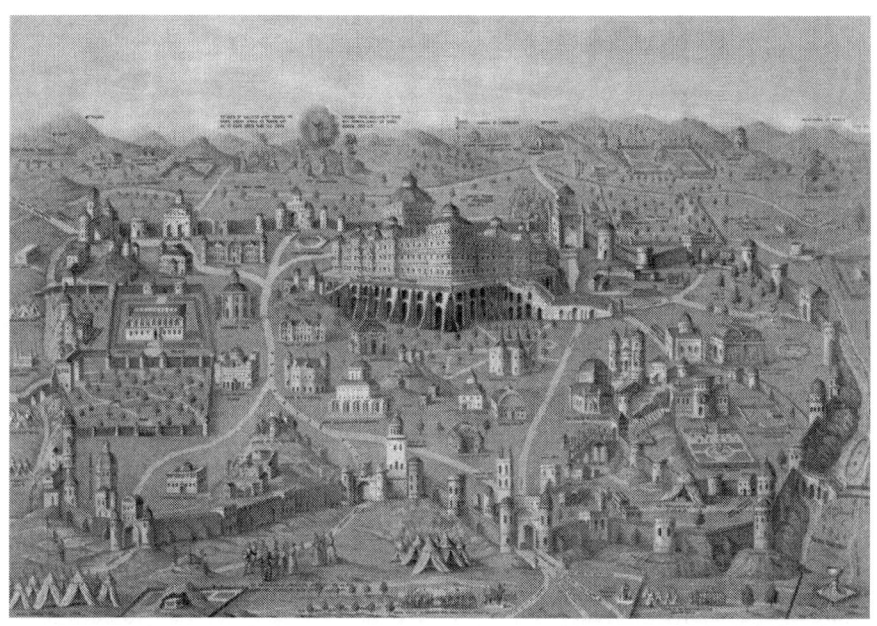

Ancient Jerusalem painting by Charles O'Donnell 1871
Library of Congress, Public Domain

Where did Robert I de Ros' wealth come from?" There was no way to make cash capitol during this time in England growing crops. Yet, he seems to have had access to extremely large sums of ready cash during his lifetime. Piers de Ros had a good visible income source, but Robert I de Ros looks to not have a source of income at all, and yet Robert I de Ros is listed as participating in the First Crusade, an expensive undertaking.

As Robert I de Ros is listed as being a Munificent Benefactor to the Knights Templar's at the very start of the Knights Templar's founding, who exactly would he have personally known to have so generously donated to this group?

No one generously donates extremely large amounts of cash unless they have connections to the organization in some manner. Robert I de Ros being with the first 9 Knights Templar in 1119 into Jerusalem as an assistant of some sort prior to the founding of the Knights Templar organization, he then must have known somebody because they were all related to one another. These first 9 Knights Templar petitioned the newly placed King of Jerusalem, Baldwin II, to form an Order to Protect the new wave of pilgrims flocking to the newly acquired Holy City of Jerusalem.

It should be noted that this "Order" of knights was made up of nine older gentlemen, hardly an adequate policing force. This new order was granted the temporary right to 25% of the Temple Mount as their base of operations in Jerusalem. Why would the King of Jerusalem hand over 25% of the Temple Mount? My research into King Baldwin II, makes him close kin to William the Conqueror's wife Matilda.

Questions of their true intentions arose after they gained approval from the King of Jerusalem Baldwin II, to excavate under the ruins of the Temple of Solomon, known today as the Temple Mount. It seems that of the founders of the Knights Templar, Hughues de Payne and Godfrey de Saint-Omer and the other 7 knights seemed to have had an idea of what exactly what was buried beneath the Temple Mount.

In Jerusalem in the year 1867 the Royal Engineers, a British archeology team, re-excavated beneath the Temple Mount and found the old Templar tunnels and several Templar items such as spurs, swords, and digging equipment. They verified exactly what the original 9 Knights Templar were doing during the first 9 years in Jerusalem, they were not policing the surrounding area and protecting the pilgrims as first stated. Modern day scholars have always questioned how an order of just nine middle aged men, could have been an adequate police force in the barbaric lands surrounding Jerusalem. Why did they dig tunnels under the Temple Mount? They seemed to know exactly what they were looking for and where to find it? Did they have insider information of what was buried under the Temple Mount? No one knows for sure what the 9 Original Knights Templar found beneath the Temple Mount.

Some believe they found the treasures that the Jewish leaders hid just prior to the Temple Mount being destroyed by the Roman Military in 69 A.D. One of the Copper Scrolls that was uncovered at Qumran in 1947-1956, gives an account of what exactly the Jewish leaders hid. A vast amount of treasure had been buried under the Temple Mount.

The famous Copper Scroll of Qumran, that was uncovered at Qumran in Israel 1947-1956, gives an inventory of these treasures, tons of gold and silver, multiple religious relics and a list of other scrolls that were hidden as well. Scholars have tried to decipher the Copper Scroll.

Consensus puts the totals around:

26 Tons of Gold Total

65 Tons of silver

Total Located in 60 different sites listed on the Copper Scrolls, plus various precious vessels, and ornamentation.

The Copper Scroll listed a combined value of over $2 billion dollars' worth of treasure. Scholars agree that the task and expense of engraving a list upon copper scrolls is extremely challenging. Meaning who would invest the time or energy into doing a hard task without having a genuine need to have a record that can stand the test of time. The Jews were aware that the Roman Legions were coming, put down the insurrection in Jerusalem. An occurrence they felt could wipe most of them out. They needed a record of where the Jewish Treasures were hidden. The Qumran Copper Scrolls contains: the 60 plus sites of the treasure of Jerusalem hidden in 69 AD, further description on locations, how to dig or locate each site, distance described in cubits (a cubit is 18 inches), treasure description located per site.

Copies of the copper scrolls were hidden around Israel prior to the invasion of the Romans in 70 CE, so that in the future after the Romans had left the remaining Jews could then locate the hidden treasures. No story or tale is imprinted on the copper scrolls just totals and locations.

During the First Crusade 1096-1099 could the Normandy Knights have obtained a Copper Scroll or two with the site lists of the Jerusalem area treasures, that they could come back to in the future?

Say 9 guys with their support staff of 10 men each? The Qumran site is located on a dry marl plateau about 1.5 km from the northwestern shore of the Dead Sea, near the Israeli settlement and kibbutz of Kalya. The First Crusade was in and around this area for roughly 2.5 years.

The original 9 Knights Templar and their assistants dug under the Temple Mount for 9 years after which time they loaded up huge wagons with tarp covers. King Baldwin II provided an armed escort out of Jerusalem. Well my guess they found an exceptionally large amount of something.

In 1129 de Trusbut, de Ros and de Mowbray donate to the new organization of the Knights Templar in England with the Pope's sanctioning. The Pope received 10% of whatever was Declared by the Templars in the Middle East. Since Robert I de Ros was known to have participated in the First Crusade and the Excursion to the Holy Lands by the first 9 Knights Templar. With just what exactly and how much of it, did Robert I de Ros return with? He built the Kirkham Abbey and he is listed as the Knights Templar's munificent benefactor.

My suspicion then is, Robert I de Ros returned from the Middle East after the First Crusade or on his subsequent trip to the Holy Lands with copious amounts of Spoils of War. Spoils of War, taken during any of the Crusades were a non-taxable occurrence designated by the Catholic Popes. Robert I de Ros married Sibylla de Valoines, and by her had son named Everard de Ros. When Robert I de Ros dies Sibylla deValoines next marries William de Percy, and upon his death she marries Ralph D'Albini. Point of Fact, all three of Sibylla's husbands go on to be grandfathers of 3 different Magna Carta Surety Barons. That seems odd, 3 family connections to the Magna Carta Barons. I considered the first 3 familial connections of the 25 Magna Carta Surety Barons stunning. The entire list of 25 Barons are all multiple times relatives of each other. I have an accounting in a later chapter.

But first what is a Munificent Benefactor? Being of humble birth, I had no context in my life to even understand this fine word.

Munificent Translation: mu·nif·i·cent / myoōˈnifəsənt, An adjective: (Of a gift or sum of money) larger or more generous than is usual or necessary.

That is way more than I expected. I had essentially walked into this history of the Rose Family blind. With no preconceived thoughts or ideas as to what I would find only that I wanted to find the "Power and the Control" of the Rose Family. Everard de Ros the first-born son of Robert I de Ros, married Roesia the daughter of William Trusbut, who is son of Gaufred FitzPagen, brother of Hugh de Payens.

From wikipedia [1]: "Fitz is a prefix in patronymic surnames of Anglo-Norman origin. This usage derives from the Norman fiz / filz, pronunciation: /fits/ (cognate with French fils < Latin filius), meaning "son of"... Later in the 1600's the term Fitz would come to mean "Bastard son of" but not at this time in the early 12th Century.

After the deaths of Rosia's 3 brothers, Roesia deTrusbut wife of Everard de Ros, was one of the co-heirs of William deTrusbut her father, together with her sisters Hillaria and Agatha de Trusbut. Roesia de Trusbut was not only heir to her father William de Trusbut/FitzPagen/de Payne Estate, but also her mother's father's estate as well. Her mother was named Albreda de Harcurt, daughter of Roesia's Grandmother Roysia (not Roesia), one of the daughters and co-heirs of Pagan Peverell of the de Payens family, son of the illegitimate son Ralph of William the Conqueror, who was the Standard-Bearer to Robert Curt-hose, William's legitimate eldest son, during the First Crusade to the Holy Land.

Did you just read the name Pagan again? Thus, making a two way, possibly a three-way cross into the Trusbut/de Payne/Pagan family by the de Ros family. Roesia deTrusbut-de Ros's two sisters, Hillaria and Agatha failing of issue (no children), leave their posterity to Roesia. She then passes the de Trusbut or FitzPagan Barony to Lord Everard de Ros her husband, who then becomes Baron of Trusbut by specific warrant from her father William deTrusbut/FitzPagan upon his death.

With that also comes the Heraldry Shield of deTrusbut, with the "Trois Bouts d'Leau" on the shield. At this point in time in history there were no regulations yet on Heraldry inheritance. No regulations to stopping the ability of a Peer, on the passing of his Heraldry to someone other than his own first-born son. Meaning, it could be given at will. This being a very important point about Hugo de Ros/Rose the First Baron of Kilravock' own Heraldry Shield containing the Trois Bouts d'Leau.

Everard and Roesia de Ros's first son and heir is Robert II de Ros aka Furfan, who married Isabell MacWilliam, a daughter of the King of Scotland, William the Lion. Robert with Isabell MacWilliam had two sons. William and Robert de Ros. Robert married a Scottish King's, daughter? Start thinking of the consequences of such a match. I will discuss this in a later chapter.

Robert goes on to build the Castles of Helmesly/Hamlake and Warte/Wark/ Warke and gave to the Knights Templars the Preceptory of Rievaulx Abbey and Ribstan Abbey, in Yorkshire, England. To his son, William (King William's namesake,) he gave the Castle of Helmesly/Hamlake, with the Appurtenances and the Advowsons of the Monasteries of Kirkham, Rievaulx, and Warden, To son William, a Barony in Scotland, to be held by William, and his Heirs, by Knight's Service to the King (of England).

This then is the line that the Roses will not cross in the future as the Barony in Scotland hinges upon "Knights Service to the Kings of England," not Scotland. Important to the Roses of Scotland as the first-born son of Robert II de Ros William gets the Barony to be handed down through first son.

To his second son Robert the Castle of Warte/Warke/Werke with the Appurtenances to them.

Trois Bouts d'Leau

Chapter 11

Robert II "Furfan de Ros seems to be the connection I was looking for in the search of the ancestors to Hugo de Ros/Rose the First, Baron of Kilravock Scotland.

Trois Bouts d'Leau photographed by author Susan Rose 2019

The Shield on the left shows the Gold background that designates Royal. Prior to Robert marrying Isabella MacWilliam daughter of King William the Lion, the shield background was Red, shown on right. Same Heraldry two different versions.

Hugo de Ros/Rose the First, Baron of Kilravock's Heraldry Shield

Trois Bouts d'Leau Heraldry on Kilravock Castle Scotland Photographed by author Susan Rose 2019

The use of another knight's heraldry was a capital offense in those bygone days. It was not done if you wanted to stay alive. Heraldry shields all meant something. I researched into exactly what was on Robert's Shield. French Heraldry definitions showed in French language that Trois means three, that is easy to see as there are three curly things on the shield. "Bouts d'Leau" means buckets/bags of water that are thrown over a saddle horn to carry water through the desert. Robert's Heraldry shows a set of three, two at the top and one lower. They are what we now call canteens. Only one family gets to use that set up in exactly that configuration, period. So, what does the Trois Bouts d'Leau represent?

A Container of Water Is for us a Symbol.

Not of affluence but of Order.

Of control over the Uncontrollable.

This container of Water,

Is useful and as Such,

Infinitely soothing to the Eye.

Yes, that makes sense, if you are following a guy into battle in the desert. Follow the guy with the control over the uncontrollable. It is an analogy for Power and Control. Showing Control over the Uncontrollable. We as humans know very clearly water goes wherever it wants when it wants. So then that being said, a guy who can control water is the guy with All the Power.

I needed to know more about Robert. Why was he being referred to as "Furfan," a nick name. It seems that in Old English Furfan/Fursan means "Halo or Aura" or more directly a "Fuzzy Fan" (fuz fan) to (fus fan) to (fur fan) see the genesis of that term? "F" s and "S" s are interchangeable in Old English. Two different spellings for the same word. Standardization of spelling of the English language does not come about for another few hundred years.

A fuzzy fan over Robert's head, a Halo or Aura attached to him. The English populous made up terms for what they are trying to convey, exactly like our slang words of today. Well then Robert seems to be at first glance to be quite fun, a glowingly nice, smart kind of guy with huge family connections. So that on formal documents the name "Furfan" is all that is needed to identify him, means everyone knew who he was. The Aura around him that is being referred to, means he had a presence. A get out of his way he is an important person type of Aura. I in my personal life have met those important people, you feel it in their presence. They are "That Someone," and "Don't mess with Them!" feeling over comes you when they are near. This is an instinct in humans, you know that prickly feeling you get in the back of your neck when danger is around? Robert must have been very much respected in the least, or at the most a very Influential Power Broker to the Kings.

A Catherine's Wheel

Chapter 12

What was the Heraldry Shield of the de Ros's prior to William de Trusbut/FitzPayen giving the Edward de Ros, the Trois Bouts d'Leau Heraldry Shield? Researched into early Heraldry of French Normans, shows the 5 de Ros's Knights carrying a Catherine's Wheel as the Adornment on their Heraldry Shield Flags.

What is a Heraldry Shield used for? The start of Heraldry Shields most likely started with medieval knights wanting to help their allies or enemies to distinguish them from others on the field of battle. This heraldry often contained very personal devises painted onto these Shields and Banners, distinguishing the carrier immediately, sort of like our todays sports team's shirt Logo system.

I researched the regulations of Heraldry in Scotland, England, and France. Heraldic Shields are first regulated in Scotland starting in year 1295. I found this date to be of great coincidence, considering Hugo de Ros/Rose the First seemingly has no parents listed, but uses the incredibly famous de Ros, Trois Bouts d'Leau Heraldry in 1280. I do not believe in Coincidences…only Realities. In Scotland after 1295 a Heraldry Shield could only be passed first son to first son, period. Not to third or fourth sons…coincidence? No.

A Catherine's Wheel was not a cleated wagon wheel for use in the mud as I had first thought when I viewed it, but it was a horrific instrument of torture.

Martyrdom of St George, Collegiate Church of St. Georg in Tubingen, Alamy

Why is it named a Catherine's Wheel? Definition of a The Catherine's Wheel is: "As one that runs rampant." A popular firework of today is named "A Catherine's Wheel," this firework runs out of control seemingly exploding without regulation or sense.

It is named for a Catholic martyr named Catherine of Alexandria, who lived during the fourth century A.D., who was by all accounts an uncommonly well-educated woman. She participated in multiple debates with the male leaders of the Alexandria society, not previously done before by a woman. Catherine of Alexandria successfully changed the minds of many of the highly placed leaders of this early branch of Christianity, the Catholics. This invasion into the male leadership, was not looked upon favorably by the other leading Catholics of the time. They condemned her to be "Broken on the Wheel."

Here is a description of what that involved. Death occurred after having one's arms and legs broken, then wound through the wheel. The victim, then being tortured for days until death finally occurred.

Now when Catherine touched the wheel to be used in her torture, the wheel was reported to have miraculously flown into pieces. That occurrence seems to not have deterred the leaders of this new Catholic division of Christianity, so they beheaded Catherine instead. Then the assumption is; that the holders of the Catherine's Wheel as a Heraldry Shield also known as the deRos family, means that they were the guys who implemented the use of such a device, the afore mentioned Catherine's Wheel, "Torture System Deluxe."

I am sure that Everard de Ros was incredibly pleased that the de Trusbut/FitzPaynes, wanted the de Ros family to now carry the "Nice Guy," Trois Bouts d'Leau Heraldry. The de Trusbut/FitzPaynes wanted the de Ros's covered by the Protection of the de Paynes. This Trois Bouts d'Leau Heraldry Shield now gave them a Shield to Hide Behind. Clean up their image, so to speak. A "Politically Corrected", a more important shield… not just a, "Terrifyingly Frightening Heraldry Shield."

The Trois Bouts d'Leau was officially transferred over to this specific branch line of the de Ros family tree in year 1170AD, the year of Robert II's birth. So, it stops at that point any other possible de Ros line confusion. Meaning the Scottish line of Hugo de Ros/Rose the First, comes from this de Ros family branch and no other. The other 4 de Ros brothers' lines continued using the Catherine's Wheel as their Heraldry.

Great Scrolls of the Board of Normandy
Chapter 13

"The land of "William Pevrel of Nottingham" is confiscated and he is "Dis-Inherited" by Proclamation of King Henry II for reason of attempted poisoning by drink of Ranulph, the Earl of the County of Chester, Vicomte de Bayeux, Vicomte d'Avranches." Quote from the Great Scrolls of Normandy, which was part of England at the time but pertained to happenings in Normandy.

Multiple accomplices are noted by their holdings not their names. Most notable to me is the holding of Bonneville-sur-Touques Castle, at that time was held under the name of Edward de Ros, son of Robert I de Ros. So, in context who exactly is this William Pevrel? And why was Edward de Ros acting as his accomplice? Please remember standardization of spelling does not come into existence until the 1400's. Pevrel, Payen, de Payen same name. Which makes William Pevrel a cousin to Edward de Ros by way of de Trusbut AKA FitzPayen AKA de Payen.

Now let us work on the "Dis-Inherited" portion of the above proclamation. To be dis-inherited means you are family to the person dis-inheriting you. So, digging further it seems William the Conqueror did have an illegitimate son by the name of Ralph. He had no last name listed to a holding as he was illegitimate, Ralph would not have a legitimate holding alliance name.

As a child Ralph was "given" by William the Conqueror to the family of de Payen to raise. In that same family his stepbrothers were Hugh de Payen third born natural son (originator of the Knights Templar) and William FitzPayen AKA William Trusbut first born natural son. Remember de Trusbut is a descriptive name of Tres Beau de L'Eau being the Heraldry Flag of William FitzPayen AKA Trusbut.

The de Ros family married into the Trusbut family by marrying the first daughter of William Trusbut. Trusbut's son was killed is a horse-riding accident. Trusbut transferred his Flag/Heraldry to Edward de Ros. It is still to this day the Heraldry to the Clan Chief of Rose and the Duke of Rutland.

So, the connection to the de Payen/Pevrel family and the de Ros family is very tight. Meaning Hugh de Payen (originator of the Knights Templar) was Edward de Ros's uncle by birth. Thus, making William Pevrel a Step Cousin to Edward de Ros. William Pevrel son of Faulk Pevrel, son of Pagan, son Ralph the illegitimate son of William the Conqueror.

Historian Thomas Stapleton spent years going through the very same documents I have been accessing, to the same conclusion. To Dis-Inherit a Relative of Henry II, that person must be a Relative of Henry II. William Pevrel was William the Conqueror's Illegitimate Great Grandson a Natural Cousin to King Henry II.

This is now making sense as to why the first listed benefactors to The Order of Knights Templar were William Trusbut and Robert I de Ros in 1129.

Now pay close attention, William Peveral's land of Nottingham is confiscated by Henry II. It is then turned over to Albreda de Harcurt, daughter of Roesia de Trusbut's Grandmother Roysia, one of the daughters and co-heirs of Pagan Peverel. Nottingham now belongs to the Robert II de Ros, as a gift from Roesia de Trusbut his mother. The Belvoir Castle is in the Nottingham area of Old. It is still held by The Duke of Rutland (a de Ros descendant) today. This now places Robert II de Ros in Nottingham as "The Sheriff of Nottingham" while he was acting Sheriff of Cumberland during the original time of Robin Hood.

Married to a Kings Daughter
Chapter 14

The first-born son of Everard de Ros was:

Robert II "Furfan" de Ros born in 1170 AD, who married Isabel MacWilliam, natural child (meaning illegitimate) daughter of William the Lion, the King of Scotland. Robert was married to a Scottish King's daughter! Remember, I was trying to figure out where the Power and Control came from? I am finding Power and Control in literally everyplace I look now!

" William, I of Scots (from the University of Mull) Known as William the Lion – though not in his lifetime - perhaps because his Heraldry Shield bore the 'lion rampant' that became the royal coat of arms in Scotland): born about 1142 or 3; succeeded his brother Malcolm IV in 1165. Married Ermengarde de Beaumont, a granddaughter, through an illegitimate daughter, of Henry I of England in 1186. Died 1214; succeeded by his son Alexander II. Created Earl of Northumberland, William the Lion lost his estates when his brother Malcolm IV surrendered what are now the northern counties of England to Henry II of England in 1157. His reign saw many attempts to repossess them, including the war of 1173-74, ended in William's capture at the Battle of Alnwick, as a result of which Henry II became overlord of Scotland. King William the Lion was housed in Falaise Normandy during his incarceration by Henry II, he was freed with the Treaty of Falaise. This was set aside under Richard I, The Lionhearted, who sold sovereignty back to William to fund his crusading.

William the Lion founded Arbroath Abbey 1178, which may show the way he strengthened the law, the peace, and the civilization (in a more European sense) of Scotland, where the people were still mostly Celtic.

In early 1190, in Haddington, East Lothian, Scotland, Robert married Isabella Mac William (Isibéal nic Uilliam), widow of Knight

Robert III deBrus. Isabella was the illegitimate daughter of William the Lion, King of Scots by the daughter of Richard Avenel. Isabella MacWilliam was previously married to Robert III de Brus."

De Brus, say that out loud, "de Brus..." That name sounds like Robert the Bruce, King of Scotland; well yes, it is, the very same family. The de Brus family was also a Norman Knight Family. So, then looking forward in time, Robert II de Ros was married to King Robert the Bruce's Great Uncle's widow. Follow that?

"William the Lion was housed in Falaise Normandy".

King William the Lion eventually married Ermengarde de Beaumont, a granddaughter, through an illegitimate daughter of Henry I of England, son of William the Conqueror. Please then note, that both William the Lion and Robert II de Ros are from one huge extended family. King William the Lion was the Earl of Northumberland. Didn't Piers de Ros get Northumberland in his marriage to Adeline d'Espec? Land and castles were owned by the "current" King of England? Control over them was given out on his whim. So, then this puts William the Lion's daughter Isabella MacWilliams, in roughly the same location on the map, as Robert. Such a prize as she, could not be left alone, so of course Robert married her, as he was that kind of guy.

Ten years later in 1200, Robert was to escort King William the Lion to appear before King John of England. Re-visit the quote from the above info on King William the Lion: "William the Lion introduced Norman families, like deBrus's and Stewarts, into Scotland and he founded Arbroath Abbey 1178, which may show the way he strengthened the law, the peace and the civilization (in a more European sense) of Scotland, where the people were still mostly Celtic.

Children of King William the Lion

Chapter 15

With wife: Ermengarde de Beaumont (great granddaughter of King Henry I) Edinburgh Castle was her dowry.

Margaret (1193-1259) married Hubert de Burgh 1st Earl of Kent Justiciar of England and Ireland for King John and Henry III, Herbert was cousin to Robert II de Ros by way of Serlo de Ros/de Burgh (of the 5 de Ros Norman Knight brothers 1066) The de Burgh line of Ireland, descends from Serlo de Ros/de Burgh

Isabel (1195-1253) married Roger Bigod, 4th Earl of Norfolk, Surety Knight of the Magna Carta

Alexander II of Scotland (1198-1249)

Marjorie (1200-1244) married Gilbert Marshal, 4th Earl of Pembroke whose father was William Marshall Magna Carta Surety. Interred to the left of William Marshal at Temple Church London

By unnamed daughter of Adam de Hythus:

Margaret, Married Eustace de Vesci, Lord of Alnwick, a Magna Carta Surety

By Isabel de Avenel:

Robert of London

Henry de Galightly, (son Patrick was competitor for the crown in 1291)

Ada Fitzwilliam, 1164-1200 married Patrick I, Earl of Dunbar

Aufrica, married William de Say, great great-grandson competitor for the crown 1291

Isabella MacWilliam born 1165 first married Robert III de Brus of Annandale, great great-Uncle of King Robert the Bruce (7th of Annandale). Then **Robert II de Ros** is a Magna Carta Surety interred to the right hand of William Marshall in Temple Church London. His great great-grandson competitor for the crown 1291.

Hubert de Burgh

Chapter 16

Property of Belvoir Castle Archives Title to the Scroll
"Hugh de Burgh Genealogical Connections to De Ros"

Photographed by author Susan Rose Scroll property of Belvoir Castle Archives, "Hugh de Burgh Genealogical Connections to De Ros" Photographed by author Susan Rose 2019

I have read many documents from Castles, Manor Houses and Religious Houses. Most all I can read and draw my own conclusions. One stood out as needing of more attention. A scroll showing the family connection between the de Ros family and Hubert de Burgh located at Belvoir Castle, England. I contacted the secretary of David Manners, 11th Duke of Rutland current owner of that Castle. I asked if I could possibly view that scroll. Amazingly within 3 hours I was contacted back by the Archivist of Belvoir, offering a date to view it.

Of course, I went to view and photograph the de Burgh Scroll. To see visible proof in living color that what I have been researching for 10 years was in fact real and not my imagination. What it shows is the family tree with Heraldry Shields of how the de Burgh line flows from the de Ros line. Then it goes on from there. The Scroll is in amazing shape. The paint is still vivid, rich Gold and Red. The Heraldry is shown with the supporters all in full color, 20 feet long. Standing with the Duke, and his Archivist talking about that Scroll, I will never forget. Amazing.

The Archivist pulled another scroll that he thought that I would like to view. A smallish scroll with what we refer to as chicken scratches and lines. On that scroll are the names of the competitors for the Crown of Scotland year 1291 and how they all descended from King William the Lion. Document property of Belvoir Castle Archives

"Competitors for the Crown 1291,"

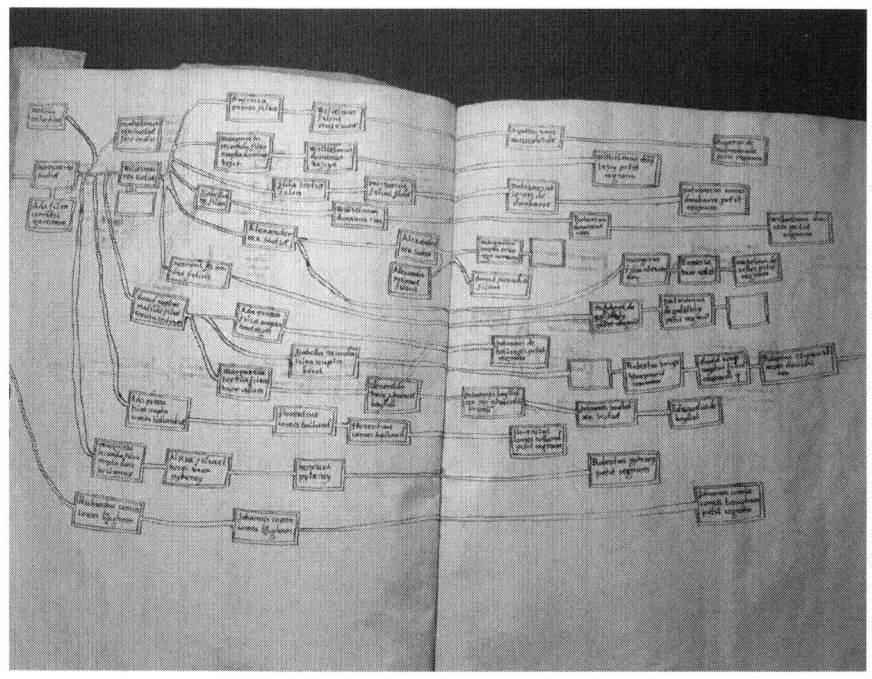

photographed be author Susan Rose 2019

William de Ros the 3rd great grandson of Robert II de Ros was on the above Document. Hand over mouth, trying valiantly not to scream out loud in front of the Duke of Rutland. I did manage that, though I did a bit of screaming later on the way down the hill in the car from Belvoir Castle.

The archivist gave me a private tour of Belvoir Castle (yes one of Robert's Castles back in the day). There I came face to face with a Holbein painting of Henry the VIII, I could have reached out and touched Henry. The 11th Duke of Rutland, David Manners is also a descendant of Robert II de Ros. He to uses the Trois Bouts d'Leau Heraldry as his own.

The Scottish Roses have changed their name from de Ros to Rose at the time of the Knights Templar Trials, very possibly a way to Hide in Plain Sight. The Scottish Rose Heraldry was amended in the 1400' to add a Boar's Head in the Center of the Trois Bouts d'Leau.

Back to Hubert de Burgh, he was the Justiciar of England and Ireland under King John and Henry III. One of the most influential people in England of the day. I am going to lay out some interesting facts:

Hubert de Burgh:

Cousin to the Robert II de Ros

Born 1170

Married to a daughter of King William the Lion

Regent to Henry III, son of King John

Robert II de Ros:

Cousin to the Hubert de Burgh

Born 1170

Married to a daughter of King William the Lion

Raised Henry III, son of King John

There are no such things as Coincidences, they are Realities you were not aware of.

3rd Crusade with Richard the Lionhearted

Chapter 17

Having uncovered exactly who Robert II de Ros was, now I am now going to address what he did in his 57 years. The 3rd Crusade with Richard the Lionhearted, started in 1190 AD. Accompanying King Richard are a contingent of Yorkshire Knights. Robert II de Ros one of the Knights Templar on the 3rd Crusade, was also a member of the Yorkshire Knight Contingent.

Robert's paperwork gifting his own resources to the Knights Templar organization making him a fighting Knight Templar still exist. Robert's membership as a Fighting Knight in the Order was later rescinded down making him an Associate Knights Templar after he returned from the 3rd Crusade.

Richard the Lionhearted by Merry-Joseph Blondel located in the Palace of Versailles, France Public Domain

On the way home from the Crusade, King Richard the Lionhearted was captured and held for ransom by the King of Austria and held for 4 years until his "King's Ransom" (yes, this is where that term originated) could be raised. In other history books about that same time, Robert was soundly criticized for returning to England ahead of the rest of the Crusaders.

I considered what Robert did upon his arrival back to England. Researching the History of Rievaulx Abbey, it is noted that the Rievaulx Abbey along with 2 other Abbeys in one year raised 25% of the Kings Ransom. At that time though Rievaulx Abbey was under the control of Robert II de Ros, who later would gift that Abbey to the Knights Templar organization. The crop production of Rievaulx Abbey that paid 25% of the Kings Ransom was in Wool. But why would Robert be so inclined to raise so much of the Kings Ransom himself? I can only guess, but...

Reportedly there seems to have been a bit of a "Dust Up in Acre" over the King of Austria's Flag being torn down while flying over Acre, by the Yorkshire Knight Contingent. Then that flag being thrown down into the nasty muddy moat below. Robert II de Ros was head of that same Yorkshire Contingent, he was probably connected in some way to the Austrian King's Flag debacle and by that subsequently responsible for having King Richard being kidnapped and held for ransom by the less than amused King of Austria. Now skipping ahead to 1197, the 8th year of King Richard the Lionhearted reign.

King Richard had been free from Austria for 3 years at that point in time. Robert was again with King Richard in Normandy France, now for what reason it is not entirely known and just what transpired preceding this next occurrence is now anyone's guess. But it was recorded in a local tableau of the time. Just that Hugh de Chaumont was committed to the custody of Robert, at Bonneville sur Touques Castle. Remember the de Ros family are the Castellans of this castle.

Robert then in turn entrusting this new prisoner to his own man, William de Spiney. The latter allowed de Chaumont to escape out of the Castle at Bonneville-sur-Touques. King Richard the Lionhearted, thereupon hanged de Spiney and collected a fine of twelve hundred marks, or roughly $1,752,000. In today's dollars from Robert but allowing Robert to stay alive. He was worth much more alive than dead to King Richard.

Oh, lots of problems in that last paragraph. This is where my exhaustive research pays off. This Castle Bonneville sur Touques is first seen in the English Tax Roles being attributed to Piers de Ros' father Guilliume de Rots/de Ros. It being a hereditary Barony held in trade for service as a knight to the King. It was inherited first son to first son, through this branch of the de Ros Family.

So, then who was the prisoner Hugh de Chaumont?

Hugh de PAYENS (PAYNS); 1st Grandmaster (and founder) of Knights Templar Wives: Mathilde (Maud) & Catherine ST. CLAIR

Possible Children:

#1 Thomas de Payne;

#2 Marguerite de Gisores;

#3 **Hugh II de Payens de Chaumont**

Making Hugh de Chaumont a Great Grandson of Hugh de Payens, Grand Master of the Knights Templar.

Then that makes Robert and Hugh de Chaumont cousins, by way of the de Payens family line.

"AD 1196 IMPRISONMENT OF HUGH DE CHAUMONT: Of the capture of Hugh de Chaumont In the same year a battle was fought between the followers of the French and English kings in which Hugh de Chaumont a great friend of the former monarch was taken prisoner and brought before the king of the English who gave him into the custody of Robert II de Ros that knight delivered him to the care of William de Spiney an attendant of his owing to whose treachery he escaped for he obtained the permission of the aforesaid William and let himself down from the wall of the castle of Bonneville sur Touques where he was confined and thus took his leave of them. The king of England was greatly enraged against Robert de Ros for this and took from him a thousand two hundred marks of silver for his offence and ordered William de Spiney to be hung on a gibbet." Roger of Wendover's Flowers of History: Comprising the History of England. (Public Domain)

Robert was possibly the wrong guy to put in charge of so important a prisoner as this Hugh de Chaumont, his own cousin. For whatever reason, he was imprisoned. All this occurs after Richard the Lionhearted was ransomed for the price of 150,000 marks, or the equivalent value of $219,000,000. In today's dollars. A vast amount back then.

Rievaulx Abbey was renowned for its wool production. Wool of this quality did not originate on English native sheep. Sheep of this quality were very new to England; they had only been indigenous prior to this in the Middle East. These sheep were referred to in that time as "Walking Gold Sheep". Meaning a pound of Wool equaled the same value as a pound of Gold, literally. These "Walking Gold" sheep arrived into England sometime shortly after the First Crusade. Meaning this family, the de Trusbut/FitzPayen who is specifically noted through its Heraldry as being on the First Crusade, more than possibly brought back these fine sheep with them on their return from the Holy Land.

Remember Robert I de Ros, who was also a munificent donor to the new order of Knights Templar? He too had been on the first Crusade to the Middle East. These guys living in the same area together and going on the First Crusade together and being related to each other. Brought a few extra things back with them during their travels to the Middle East. One of those things, being these same Walking Gold Sheep that Rievaulx Abbey farmed.

Yorkshire Contingent

Chapter 18

"When York's stout men were chosen as the Best of All the Land. To grasp a Soldiers Glory, in the Breach where Death Will Stand."

Below is a list of the powerful Yorkshire Contingent from the 3rd Crusade with Richard the Lion Heart. The men listed here were stand out soldiers for King Richard. I have found mentions of this "Yorkshire Contingent" when researching this book, but no listing for the actual names previously. I could find that Robert II de Ros was a leading member, but no one had mentioned the other names, until now.

Named Members of the Yorkshire Contingent:

Sir Robert de Turnham

Stuteville of Kirby Moorside

Ralph de Granville of Coverham (who died at Acre)

Roger de Granville of Coverham (led the desperate but unsuccessful assault of Acre on 30thNovember, 1190)

The De Ros of Hamlac/Helmsley

Warin FitzGerald of Harewood

Knights of Mara

Ralph Teisson (the Badger)

Gerard de Furnival of Sheffield

Ralph de Tilli of Doncaster (cousin to The De Ros)

Humphrey de Veilli

Robert de la Lande (probably of the Micklefield or Harewood Family)

During the 3rd Crusade a battle is described with reference to the actions of the Yorkshire Contingent. The Austrians and the English threw up their ladders over the ditches, and the infidels coming out from the city carried the ladders away from the Austrians. But Ralph de Tilli roped the ladders back from the infidels and 3 of the Contingent ascended the ladders and extinguished the Greek Fire thrown by the infidels.

During this fight, the Austrian men put up the King of Austria's Flag. The 3 men of the Yorkshire Contingent pulled the Austrian Flag down and tossed it into the Cesspit Ditch. to the Yorkshire Contingent, this was their Victory alone.

This action goes on the be the cause of the capture and being held for ransom of King Richard the Lion Heart by the Austrian King while on his return trip. Which is the foundation for the phrase "Worth A King's Ransom."

Richard the Lion Heart's "King's Ransom" was the value of 4 years' worth of England's entire GDP. Robert II de Ros (The De Ros) raised 25% of that Kings Ransom with the organization of the wool from 3 Abbeys one of which was Rievaulx Abbey.

My wonder is, why would Robert II de Ros assume 25% of the debt for the Kings Ransom? My guess he might have been one of the 3 who threw the Austrian Flag in the cesspit along with his cousin de Tilli. Conjecture Only.

"In Yorkshire, especially during the reign of the Plantagenets, they are not merely the units but the whole." ~Wheater.

The Sheriff of Nottingham

Chapter 19

The original setting of the incredibly famous story of Robin Hood, when it was first written down in 1420 was in the District of Cumberland. The Robbin Hood story prior to 1420 was only being told by story tellers, who passed on the story verbally.

Now Robert had been appointed to "police" the district of Cumberland, to start a normal 2-year term appointed by King John. Yes, the very same Evil Prince John in the Robbin Hood Story, who was in control of England as a substitute ruler, while his brother King Richard the Lion-Hearted is away fighting a war, but then in 1214 King John is on the throne of England as his older brother King Richard had been killed in Normandy, France.

One halfway through Robert's 1st year of a 2-year tenure as Sheriff, he rode away to join the First Barons Revolt in 1214 that led to the Magna Carta. Exactly what was the impetus for his leaving of this very lucrative appointed post by King John? I can only guess.

As Sheriff of the District of Cumberland, Robert would have had to enforce the rules of King John, at this point in 1214, he was angry enough to join an insurrection, which is still felt Around the World to this Day. Thank You, Robert II Furfan de Ros.

Think back on that same children's story.

The term Sheriff of Nottingham is used in that story. As French was the language of the aristocracy. The use of the word "de" means "of" or "from" not "the." Then in reference to, the Sheriff of Nottingham, that is not where the actions took place. It was where the Sheriff was from, prior to taking over the position of Sheriff. So then would the Sherriff from Nottingham be more appropriate term to identify exactly who the famous Sheriff was? Nottingham is nowhere close on a map to Englewood Forrest located in Cumberland yet that is the original site of the Robbin Hood and Little John story. Look at the location of Nottingham on a map of England, and then family the territories of de Ros' all still under control of Robert, of Nottingham and they are still there today. That little piece of info will stop you dead in your tracks, it did me.

Andrew of Wyntoun's Orygynale (Original) Chronicle (written down c.1420) places the Inglewood Forest situated in the District of Cumberland as the original setting of the Robin Hood legend, the following is taken from the chronicle

> **"Lytil Jhon and Robyne Hude**
>
> **Wayth-men ware commendyd gude**
>
> **In Yngil-wode and Barnysdale**
>
> **Thai oysyd all this tyme thare trawale."**

"Yngil-wode" is the Englewood Forrest of today located in the District of Cumberland England. You must read the above phonetically, yet again there was no standardization of written language during this time.

Look over the list of Sheriffs, there are other recognizable names on this small list as well.

1201–1203 William de Stutevill and Phus Escrar 1204–1205 Roger de Lacy

1206–1209 Roger de Lacy and Walter Marescallus 1210–1213 Hugh de Neville

1214--*********Robert de Ros** *******

1215--Alan Candebec

1216–1222 Walter Mauclerk

1223–1229 Walter Mauclerk and Robert de Vieuxpont 1230–1231 Walter Mauclerk and Thomas son of John 1232–1235 Thomas de Multon, Lord

1236–1247 William de Dacre 1248–1255 Job. Daylock

1256–1260 Wiiliam, Earl Albamarl and Remigius de Todington

1261–1265 Eustace de Balliol

1266–1269 Eustace de Balliol and Mathew of York

1270-1271 Sir Radulph de Dacre of Dacre

1272–1273 Robert de Chauncy and Mathew Cordil and Roger de Pocklington

1274–1276 Robert de Hampton

1277 John de Windeburne and Michael de Newbiggin 1278–1280 Newbiggin and Gilbert Cureweune

1281–1284 Robert de Brus

1285–1296 Sir Michael Harclay

1297–1301 William de Mulcaster

Famous Sur Names of Sheriffs District of Cumberland England in history:

de Lacy – name listed as one of the 25 Magna Carta Surety's.

de Neville – name the future famous Kingmaker during the War of the Roses.

Earl Albamarl – one of the more notable peerages in England.

de Balliol – name was to be appointed King of Scotland by Edward Longshanks.

de Brus – name of the future King of Scotland.

So, He sold a bit of English Countryside to France.
Chapter 20

Accusations have been formed against Robert II de Ros by Historians that he sold land to the Prince/Dauphin of France, to aid him in building his own personal Castles. He did sell some acres in Yorkshire to the Dauphin of France. Yes, but this acreage was not Robert's to sell in the first place, remember all land belonged to the then current King John of England. Next is the little problem of borrowing cash from the Jewish money enders of London to aid him. That too was just not to be done. But why would he do this? His Castles were already existing.

Starting an Insurrection against your King was a costly business though. Had Robert learned from the master negotiator himself, King Richard, as to how to raise money for any eventuality.

Richard the Lionhearted always his highest priority in life, was to launch his personal Crusades of the Holy Land. Crusading was vastly expensive proposition though. Everything had to be brought along with the crusading knights and their supporting staff. From horseshoes, food, all manner of armament, housing, literally everything was hauled with the crusaders. London's Jewish money lenders came to King Richard's court bearing valuable gifts, and sources of money. To aid in funding his great plans for Crusades. To overcome this lack of funding, King Richard also raised money for his crusades by selling anything he could. As in Hugh de Puiset gave 2,000 marks for the Sheriffdom of Northumberland and another 1,000 marks for buying out from attending the Crusade with King Richard. The King's own half-brother Geoffrey had to pay £3,000 for the Archbishopric of York, William Longchamp Bishop of Ely was made chancellor on payment of 3,000 marks. Burgesses (freemen) purchased their right to have their cities at fee farm (without homage or fealty to the king) varying from

£100 in the case of Northampton and 40 marks in the case of Shrewsbury.

For King Richard the Lionhearted, everything was for sale; powers, lordships, earldoms, castles, towns, and manors, literally everything. King Richard had famously said that he would sell off London to anyone who was prepared to buy it. The largest contribution for his Crusading fund came from William the Lion, King of Scotland who for 10,000 marks (£6,600 or $11,220. US) bought his release from the covenants of the Treaty of Falaise, effectively gaining Scotland's independence from England.

Smart man, Robert's father-in-law.

Here is a different angle on the sale of English land by Robert II de Ros. He sold land to the Dauphin of France, during the First Barons Revolt this is a fact. Was this to encourage the French Prince to become involved in the Barons Revolt with additional French soldiers? Was he raising money to fight King John? The First Barons Revolt eventually led to the Magna Carta. Was this a cover up yet again about the activities of Robert II de Ros?

The Dauphin was admonished by the Pope for his activities in England at this same time during the Magna Carta era. The sale of land to raise money for his already existing Castles really does not add up. Taken in context of the time frame. The selling of portions of Yorkshire was to encourage the young Prince of France to Aid and Abet Robert II, de Ros in squashing the Reign of King John then does make sense.

Robert de Ros and his Abettors
Chapter 21

The Communications from Popes,

Innocent III & Honorius III

In this Chapter are the Communications between the powers in London and these two Popes, Innocent III & Honorius III, concerning the events in the 3 years just preceding the Magna Carta to the 3 years just after. The inclusion of the names of King of Scotland William the Lion, Robert's father-in-law and Prince Lewis, son of Philip the King of France. Shows the extreme activities of these powerful English Barons in the obtaining of King John's Seal on the Magna Carta Document.

Those English Barons used the King of Scotland, Robert's father-in-law, and Prince Lewis of France, (yes that same Dauphin of France that Robert sold a few acres to) as tools to compel King John to acquiesce to their wants. The Magna Carta story in American History Classes has the story of the Magna Carta as that the English Barons compelled John King of England essentially all by themselves. To the contrary, the following letters show that these English Barons not only controlled England, but exerted influence over Scotland, Wales, France and as well as the two Popes Innocent III & Honorius III.

The unbelievable gift by King John of his Whole Kingdom of England, Scotland, Ireland, Wales and Normandy to the Pope in 1212 AD, shows that King John is aware of the extreme unrest by the English Barons at that time and his life was being threatened by them. King John looks to be enlisting the power of the Pope to help control those powerful English Barons, but King John was also being held in Temple Church during this time, why? If King John was living in

Temple Church of his own accord, then why would he be afraid for his life?

The Knights Templar should have been able to protect him, right? But then, was he there on his own accord? Were the Knights Templar assisting in holding King John against his will? As Robert was a known Knights Templar, as well as his famous family members. The implications of that fact alone are enormous.

Problems abound, the only visible power the Popes had over the English Barons at that time, is the threat of an excommunication of these same powerful Barons. That in the end seems to have been non-effective in combating those mighty 25 Barons. Most all those same Barons had participated in a Crusade of the Holy Land. Some as Knights Templar. All Crusaders were to be admitted to Heaven by a guarantee from the Popes in Rome. So, then an Excommunication by the Pope would thus be over-ridden. But then the Popes knew who these Barons were, they knew that an excommunication would not work with them. So now, these two Popes only had the power to excommunicate on paper against those very angry English Barons.

Now in actual history, the Knights Templar were the Pope's own fighting mercenary army. They then could not be used upon themselves to enforce any of the Popes wishes, as they all seem to have been intelligent men. I think they would have laughed. Look in The Rules of the Knights Templar at the back of this book. Look for the Excommunication Clause #12. A Get out of Jail Card exists for Knights Templar.

So aware that King John was being held by their own mercenary army of Knights Templar, in Temple Church London for a goodly portion of the Magna Carta time. Then who exactly were these two Popes negotiating with in these below documents? Or is it all subterfuge? What else is being hidden from these communications? The Popes could on "paper" excommunicate the Barons all they wanted. It seems to have meant nothing to the 25 English Barons. Out of all the 25 Barons Excommunicated by the Popes. Only one was ever absolved from the excommunication from the Magna Carta activities by the Pope, he was the William Marshall Earl of Pembroke, but then both Pembroke and deRos are laying side by side among the 9 Effigies in Temple Church, and they too are also listed as the same Barons who went to Rome to give the Magna Carta copies to the Pope himself.

So then if you are excommunicated, just how do you get an audience with the Pope? It seems to be more than a little misdirection going on here. Kind of like, someone did not expect their audience to add things up. They did not know that we would have the internet today to compare documents and timelines to catch incongruences in them.

Temple Church in that day was a Knight Templar affiliated Church, meaning that Robert, should not be lying in Effigy next to Marshall in Temple Church, at all. Is this the very reason famed Knights Templar Biographer, Charles Addison in 1842 miss-reported Robert's history? No, I think the missing documents by George Rose is the true answer.

After what is a very questionable death by poisoning of King John, his young son Henry III was put into the custody and the subsequent raising of, by Robert. Back in that day, who would have allowed the Excommunicated Baron, Robert II Furfan deRos to be responsible enough to raise the future King of England, Henry III? There must be way more to this whole episode in history, than what is reported in the Archives of England.

The Magna Carta Surety Barons while seemingly serving their own ends in politics, are the cause of huge collateral benefits to the general populace of England and Scotland. Then subsequently to the World at large. Thus, does the End Justify the Means in this case? I must then agree that it just might be justifiable…

Here are the communications between the Popes and London over 6 years. 3 Years prior to the sealing of the Magna Carta and 3 after:

1212. 13 May. Dover. (f. 154.) Letters patent of the King John submitting to the pope. [Opp. ed. Migne, iii. 876; Fœdera.] 15 May. (f. 154d.) Letters patent of the King John resigning his kingdom to the pope. [Opp. ed. Migne, iii. 878; Stubbs, Select Charters; Fœdera.]
15 May. Apud Templum de Well'. (f. 154d.) Letters of the King John to the pope, offering a yearly payment of 1,000 marks. [Opp. ed. Migne, iii. 881.]

1213. 2 Non. July. Lateran. (f. 154d.) Letter to the King John, thanking him for the satisfaction and submission he has made by granting his kingdom to the Roman church, from which he holds it at a yearly cess of 700 marks for England and 300 for Ireland. [Opp. ed. Migne, iii. 881.]

1213. 2 Non. July. Lateran. (f. 155.) Mandate to the earls, barons, and other great men in England to receive and obey the bishop of Tusculum, papal legate.

2 Kal. Oct. Lateran. (f. 7.) Letter to G. cardinal of St. Martin's, papal legate, authorizing him to do whatever may appear best for assisting the king and the realm of England. [Bouquet, xix. 612.] [Opp. ed. Migne, iii. 884.]

16 Kal. Oct. Lateran. (f. 6.) Mandate to the archbishop of Bordeaux and his suffragans to urge the lieges of King John in their dioceses to hasten to England for his defense, and for the maintenance of peace throughout Christendom for four years; with further mandate to them to abstain from molesting the barons in their country. Concurrent letters to the barons and lieges of the said king in Poitou and Gascony. [Bouquet, xix. 611.]

2 Non. Nov. Lateran. (f. 162.) Letter to the King John of England congratulating him on his conversion, and advising him not to deal contentiously with the prelates of his realm. Intimation to him that his envoys, John, bishop of Norwich, H. abbot of Beaulieu, R. Martel, H. de. Bova, and P. de Maulay will inform him of the pope's answer in regard to his excommunication and the interdict. [Opp. ed. Migne, iii. 922.]

2 Kal. Nov. Lateran. Mandate to the archbishop of Dublin, the bishops of Norwich and Winchester, to earls William of Salisbury, G. son of Peter of Essex; R. of Boulogne; R. of Chester; W. of Warenne; W. the marshal, of Pembroke; R. le Bigot of Norfolk; W. of Arundel; William of Ferrers; and Saer of Winchester; R. son of Roger; W. Brigerte; *R. de Ros* G. son of Ranfred; R. de Mortuomari; P. son of Herbert; and W. de Albiniaco, to complete and keep the peace between the king and the Anglican church; ordering them if any disturbance arise to do nothing against the King without the pope's advice asked and obtained. [Opp. ed. Migne, iii. 925.]

5 Kal. Nov. Lateran. (f. 163.)
Mandate to the archbishop, bishops, barons, knights, and people of England and Wales, now that peace is made between the realm and the priesthood, to remain in fealty to the King and his heirs. [Opp. ed. Migne, iii. 926; Fœdera.]

October 15. Lafford. (f. 35.) Letter from John King of England, lord of Ireland, duke of Normandy and Aquitaine, count of Anjou, to Honorius, supreme pontiff. Mindful that his kingdom is the patrimony of St. Peter, and under the protection of the Roman church, and having convened the great men of the realm, he prays the pope to take the realm and the king's heir and successor under his protection, and to grant them absolution. [Raynaldi, xx. 397.]

3 Non. Dec. St. Peter's, Rome. (f. 20.) Mandate to G. cardinal of St. Martin's, papal legate, to protect the children of the late king of

England, and to declare illegal the oaths taken by the barons to Lewis, eldest son of the king of France, and others against the late king. [Raynaldi, xx. 398.]

Concurrent letters to the bishop of Winchester, in so far as regards fealty to the king's sons; Also, with necessary verbal alterations, to those barons of England who remain in fealty to the late King (Richard); Also to the archbishop of Bordeaux, and to those barons beyond sea who remain in fealty to the king.

3 Non. Dec. St. Peter's, Rome. (f. 21d.) Mandate to W. Earl of Pembroke, marshal of England, to remain in fealty to the late King's sons.

3 Non. Dec. St. Peter's, Rome. (f. 21d.) October 15. Lafford. (f. 35.) Letter from John King of England, lord of Ireland, duke of Normandy and Aquitaine, count of Anjou, to Honorius, supreme pontiff. Mindful that his kingdom is the patrimony of St. Peter, and under the protection of the Roman church, and having convened the great men of the realm, he prays the pope to take the realm and the king's heir and successor under his protection, and to grant them absolution. [Raynaldi, xx. 397.] [Opp. ed. Migne, iii. 884.]

16 Kal. Feb. Lateran. (f. 40d.) Grant of faculties to G. cardinal of St. Martin's, papal legate in England, in addition to those already granted, of interdicting, excommunicating, and degrading prelates and others whose rebellion deserves punishment, of disposing vacant sees and abbeys to persons faithful to the king and the Roman Church, of granting indults throughout England, Scotland, and Wales; and since some clerks still adhere to *Lewis, who is excommunicate*, of depriving and *excommunicating them if within thirty days they do not withdraw themselves from him;* and also of granting dispensations to those who have taken the cross, who are faithful to the king, to return until the kingdom is settled; also of annulling the oaths of those barons and knights of England which they have taken to Lewis; and of

excommunicating the detainers of hostages faithful to the king, so that king Henry may be served and his kingdom established. [Bouquet, xix. 623.]

13 Kal. Feb. Lateran. (f. 41.) Letter of condolence to King Henry on the death of King John, and congratulation on his own coronation; commending to him the cardinal legate, whose advice he will do well to follow. [Bouquet, xix. 626.]

16 Kal. Feb. Lateran. (f. 41d.)

Monition to William King of Scotland and his abettors, to return to their allegiance and to disregard the oaths they have taken to Lewis Prince of France.

The like to ***Robert de Ros and his abettors***.

The like to Llewelyn and his abettors.

The like to the barons of the Cinque Ports and their abettors.

The like to the earl of Warren.

The like to the earl of Clare.

The like to the earl of Arundel.

The like to the earl Roger Bigot. [Theiner, 2.]

14 Kal. Feb. Lateran. (f. 42.) Letter of monition and exhortation to W. Earl of Pembroke, justiciar of England, urging him to defend the king and realm and follow the counsels of the cardinal legate, to whom plenary powers have been given.

1217. 11 Kal. May. Lateran. (f. 98.) Monition to Philip King of France to withdraw Lewis his son from his expedition against England.

2 Non. July. Auagni. (f. 119.) Mandate to the same to examine and make necessary dispositions touching a matter about which the King,

the archbishops of Dublin and York, and the bishops of London, Winchester, Bath, and Worcester have written to the pope, praying him to remove the canons regular of Carlisle, who have publicly communicated with the disturbers of the King and realm who were *excommunicated by the pope* and the legate, and have celebrated divine offices in places under an interdict, and have also voluntarily submitted themselves to the king of Scotland, who is fighting against his liege lord, and therefore excommunicate, and at his instance presumed to elect a bishop. He is to place these canons in other regular churches, to declare null the election attempted by them, to appoint to that church persons faithful to the king, to distribute the possessions and rents of the church between the bishops and canons to be appointed, the king and the said bishops attesting that such measures would tend to tranquillity, as the church of Carlisle, being on the border, exercises much influence either for or against the king and his realm.

Id. Jan. Lateran. (f. 199.)
Letter to Lewis, eldest son of the King of the French, annulling letters obtained against him from the pope, on condition of his keeping the peace with the king of England, and doing no injury to his realm, G. cardinal legate, having mediated between them. [Bouquet, xix. 647.]

9 Kal. May. St. Peter's, Rome. (f. 85.) Indult to W. Earl of Pembroke, the king's marshal, at the king's request, as a recognition of his services, not to be held responsible to anyone else so long as he is willing to do justice to complainants before the lord of the fee about things held in fee by him, the king's right being in all cases intact.

16 Kal. July. Alatri. (f. 258. Mandate to the archbishop of Canterbury and the bishop of Salisbury, on petition of William the Marshal of England, Earl of Pendr (Pembroke), stating that the bishop elect of Norwich, at that time papal legate, and H. the justiciar, wishing to bind him more fully to the king's service, offered him one of the said king's

sisters as wife, whereto, by counsel of many bishops, earls, and barons, he assented and, setting aside many noble women who were offered him, swore to marry one of the king's sisters, and the said justiciar, as king's proctor, by command of the legate and himself, and others, took oath for the king to give him one of the said sisters. Inasmuch as some jealous persons are hindering this, the archbishop and bishop are to order the oath to be observed, if it is for the good of king and realm, and can be done without great scandal. Otherwise, the matter is to be remitted to the pope.

(This has the appearance of a bribe offer to Pembroke by the Pope.)

"This [Magna Carta] has been forced from the King. It constitutes an insult to the Holy See, a serious weakening of the royal power, a disgrace to the English nation, a danger to all Christendom, since this civil war obstructs the crusade. Therefore, We condemn the charter and forbid the King to keep it, or the barons and their supporters to make him do so,
On pain of excommunication."
                              Pope Innocent III

This is an amazing set of documents to read in order of the occurrences as they were set down. Show the power of Robert. On the date 16 Kal. Feb. Lateran. (f. 41d.) Robert deRos is listed first after the King of Scotland his father-in-law. "The like to *Robert de Ros and his abettors*."

Once again to get what is being said there, exactly what is an abettor? "To Aid and Abet a Criminal" is how that same term is used in U.S. Criminal Law. So, the Abettors were guys who worked with Robert and did what he wanted them to do. Along with his Father-in-Law, Scotland's King William the Lion and the other listed Barons as well.

As I read the above, Robert II de Ros appears as an instigator to the Baron's rebellion. In other documents from the day, also report that Robert was in control of the Magna Carta situation itself. When King John (under questionable circumstances) dies, young King Henry III is placed into Robert's custody by Marshall.

Seemingly a strange move, but more like a bow to his Power and Control. The future King of England was then being raised by the Excommunicated Rebel, Baron Robert II Furfan de Ros, who was married to also the Excommunicated Scottish King's Daughter. King William the Lion of Scotland's daughter Isabella.

Anyone see problems there? I know that these are my people now but really, I can read.

I wonder why Author Charles Addison seemed to decline any knowledge of Robert II de Ros in his famous book on the Origins of the Knights Templar. More than just a little bit of shady doings going on during the time of the Magna Charta by known members of the Knights Templar themselves.

To date I find no retraction of the Pope's Excommunication of Robert II de Ros, yet he is buried in Effigy in Temple Church. Remember the Knights Templar were under the auspices of the Pope in Rome, during this time. So, this whole situation is out of alignment. Is what was being said by the Pope, said just to placate King John of England. A make-piece gesture to get King John to capitulate to the Barons.

Were those two Popes just trying to quiet everyone down? At this point in time the Knights Templar were housing King John in Temple Church itself, in a house arrest situation. There seems to be so much left out of this story. What was not reported or exactly what was covered up? Were the Excommunications real? Or were they just for the press, so to speak? Robert II de Ros was a Knights Templar as an

Associate Member up to his death in 1226/7. All of which could not have happened if the Excommunications were real.

Is this the reason for the miss reporting of facts by Charles Addison in 1848? Or did he just never look anything up? Most of these documents were to have been housed in Temple Church as the Official Home of the English Barristers still to this day. Did the law Librarian in the 1772 whose last name very coincidently was Rose in anyway, block access to important historical information from everyone.

As George Rose being the head Law Librarian of Temple Church, could he have considered and had knowledge of Robert II de Ros during the time of the Magna Carta? That same Law Librarian named George Rose would walk by that beautiful Knights Templar Effigy depicting his very own multiple times Grandfather in Temple Church, a place where he worked every day.

Could he have not known?

A Magna Carta Surety

Chapter 22

Robert II "Furfan" de Ros, a Magna Carta Surety. He was one of the 25 Barons who agreed to enforce the Magna Carta against King John of England in 1215 AD. A Surety was an Enforcer, they made sure the laws were followed.

In America, we look at the Magna Carta as a list of rules, being like our American Constitution. So, then we Americans have the "Executive Branch" of our Government to act as American Sureties, whose sole purpose is to enforce the Laws of the United States.

So, then the Magna Carta's 25 Sureties aka Enforcers were there to enforce these new rights for the good people of England. The jobs of these Enforcers were to exert absolute Power and Control (remember those 2 words?) over King John formerly known as Prince John, in the Story of Robin Hood?

These Enforcers at first glance are "Landed Gentry Barons," with a few Earls thrown into the mix. This Surety Team of 25 Barons were not so nice and sweet though. These 25 Barons were all fighting knights, many of whom had been on a Crusade. They were far from the honorary Knights we are used to seeing on modern day TV. These were chain mailed, steel helmet wearing bad asses. To put into modern context, it is like as if a well-bred biker gang of today regulated the United States Presidential Oval Office.

We in America and our politically corrected point of view, tend to think back on Philadelphia in 1775 with fondness, just preceding our break from England. Our Founding Fathers in their frock coats harrumphing and clearing their throats, setting up the rules to run our new country using this same Magna Carta document as a reference tool. In our politically corrected American heads of today maybe a fisticuff or two, but mostly just polite gentlemen. Wrong idea in

England in 1215, as it was wrong in 1775 in the United States of America.

Nice and Sweet could not be further from the truth.

The Close Family Interconnections of the Surety Barons of the Magna Carta

Chapter 23

When I first started investigating the names of the 25 Barons of the Magna Carta. I found close family connections between Robert and a couple other Magna Carta Surety Barons. I was dumbfounded by that fact alone. When I fully explored the inter-family connections of the 25 /26 Barons, they took on the appearance of a huge extended family. It appears to me to that these 25 Barons worked together in league, to exert their power and control of England, Scotland, Wales, and France.

The initial listing of #18. A Roger de Montbegon is out of character with the group. I find it remarkably interesting that he goes missing literally and is subsequently replaced with another family insider #26. Roger de Mowbray. #10. William de Hardell is the Mayor of London but has no children, meaning that he is of no threat to this syndicate of extended family. This fact is also accounted for in the Magna Carta itself. Article #61 (look for highlight.) Was there a plan to remove him prior to the Document being Signed?

While reading the family interconnections of the Barons of the Magna Carta, please note that this is not a joke. These people were for real. These combined families of 25 Powerful Barons worked together to fight the two Popes and King John combined.

The Barons because they stayed together as a strong team, won. They have been misrepresented to us in World History Class though. As just a set of random guys working for the betterment of England. It appears that this is a complete misrepresentation of fact. This does not mean that I think that King John was fine King in his management of England. Just the opposite appears to be true.

The Barons were very right with what they accomplished, and the world is now better because of them, but the term "Collusion," screams out to me.

#1 William D'Albini. — He was a grandson of the mother of No.18 by her second husband. His step grandmother was No.22's Grandmother. His granddaughter was the wife of a grandson of No. 22. A cousin was the wife of No.19, and she was a Cousin of No. 8.

#2. Hugh Bigod. — He was the son of No. 3. His granddaughter married the half-brother of the wife of No. 4. This granddaughter's husband was also the half-brother of No.15. His grandmother was the aunt of No. 24. He was a cousin of No. 24. His wife was a sister of No. 16, and a sister of the wife of No. 5. His wife's maternal great-grandfather was a brother of a grandfather of No. 6.

#3 Roger Bigod. — He was the father of No. 2. His mother was the aunt of No. 24. His daughter was the wife of a brother of No. 24. Brother in law to No.22

#4 Henry de Bohun. — His wife was a sister of No.15, and her half-brother married a granddaughter of No. 2. His son's wife was a granddaughter of the aunt of the wife of the son of No. 9. His son married a granddaughter of the sister of the father of the wife of the son of No. 9. His grandson's mother-in-law was a sister of No. 16, whose great-grandfather's brother was the grandfather of No. 6, the father of No. 5. His grandson was the son-in- law of a sister of No. 16. His son's wife was a cousin of the wife of the son of No. 9. His son-in-law was a nephew of the mother of No. 20. His granddaughter was a sister-in-law of No. 16. His son-in-law was a cousin of No. 20. His multiple time granddaughter married No.22's multiple times granddaughter.

#5 Gilbert de Clare. — He was a son of No. 6. His son's wife was a daughter of No. 12. He was a cousin of No. 8. His wife was a sister of No. 16, and also of the wife of No. 2, and her maternal great-grandfather was a brother of the grandfather of No. 5. His great-grand daughter was the wife of the great-grandfather of No. 17.

#6. Richard de Clare. — He was father of No. 5. His paternal great-grandfather and the paternal grandfather of No. 8 were brothers. He was a cousin of Nos. 8, 19, and 20. His ancestor, the first Earl of Clare's daughter, was the wife of the great-grandfather of No. 17. His grand-father's granddaughter was the mother of No. 20. His father's sister was the wife of No. 19. His wife was a sister of No. 15. His grandfather and the maternal grandfather of No. 16 were brothers. His great-aunt married an ancestor of No. 17. His father's cousin's widow was the mother of No. 21. His wife's brother married a sister of the father-in-law of No. 21. His brother-in-law married the aunt of the wife of No. 21. His paternal grandfather was a brother of the paternal great-grandfather of No. 16. He was a cousin of No. 16. His wife's maternal grandfather was a brother of the paternal great-grandmother of No. 16. His ancestor, the first Earl of Clare's wife, was of the same family as that of the wife of No. 24. His maternal grandmother's brother was grandfather of the wife of the son of No. 21. His granddaughter was the wife of a grandson of No. 19. His paternal grandfather was a brother of the paternal grandmother of No. 24. His aunt was the mother of No. 26. His granddaughter married a grandnephew of No. 26. His mother's second husband was a cousin of No. 1. His grandfather's brother was the maternal great-grandfather of the wife of No. 2 of the mother-in-law of No. 25; of the son-in-law of No. 13 and of the second husband of a granddaughter of No. 21.

#7. John Fitz-Robert. — He was a cousin of No. 25, and of No. 12. His wife and the wife of No. 25 were cousins.

#8. Robert Fitz-Walter. — He was a cousin of Nos. 5 and 6. His paternal grandfather and the paternal great-grandfather of No. 6 were brothers. He was related by marriage to No. 9. His mother's second husband was the grandfather of No. 1. His paternal grandmother had as second husband the father of No. 21. He was a cousin of No. 15. His grandmother's nephew was a cousin of No. 21, and the husband of an aunt of the wife of No. 25.

#9. William de Fortibus. — His son married a daughter of the brother of the great-grandmother of the daughter-in-law of No. 4. His wife was a sister of No. 17. His sister was the first wife of No. 16. His mother was the widow of a nephew of the grandmother of No. 23, and this nephew's first wife was a daughter of No. 8. His aunt was the second wife of No. 16.

#10. William de Hardell. — He does not seem to have been related by blood or intermarriage to any of the Sureties. (Mayor of London)

#11. William de Huntingfield. — He does not seem to have been related by blood or intermarriage to any of the other Sureties. His great-grandfather held the manor of Huntingfield, in Suffolk, as the under-tenant of Robert Malet, the father or grandfather of No. 14.

#12. John de Lacie. — His daughter was the wife of a son of No. 5. His second wife was a daughter of the son of No. 21. His mother was the half-sister of the paternal grandmother of No. 23. His paternal grandmother was a great-aunt of No. 15. His maternal grandfather's first wife — the great-grandmother of No. 23 — was the sister of the paternal grandfather of No. 24. His widow married a brother of No. 16, a cousin of Nos. 5 and 6.

#13. William de Lanvallei. — His wife and the wife of No. 14 were cousins. His daughter's father-in-law's third wife was the widow of No. 15, and his fourth wife was a sister of the wives of Nos. 22 and 25. His daughter was the second wife of a brother of No. 16. His son-in-law was a great-grandson of the brother of the grandfather of No. 6, whose wife's grandfather was a brother of the great-grandmother of the son-in-law of No. 13.

#14. William Malet. — His wife was a cousin of the wife of No. 13. See under No. 11.

#15. Geoffrey de Mandeville. — His wife was a sister of the wife of No. 6, and their maternal grandfather's sister was the wife of a grand uncle of No. 6. He was a cousin of No. 23. His sister was the wife of No. 4. His half-brother married a granddaughter of No. 2. His great-aunt was the paternal grandmother of Nos. 12 and 23. His wife's brother married a sister of the father-in-law of No. 21. His wife's brother's wife's sister married a son of the brother of the mother of No. 21 (nephew of the mother of No. 21), who was also the nephew of the grandmother of No. 8. His maternal great-grandmother was a sister of the grandfather of No. 24. His grandfather was also the grandfather of No. 23. His wife's maternal grandfather was the brother of the maternal great-grandmother of No. 16.

#16. William Marshall, Jr. — One of his sisters was the mother-in-law of a grandson of No. 4 another was the wife of No.5 another was the wife of No. 2; another was the mother-in-law of No. 25, and her husband's second wife was a granddaughter of No. 21. He was a cousin of No. 6. One of his brothers married a granddaughter of No. 4; another married, first, a sister of the wives of Nos. 22 and 25, and married, secondly, a daughter of No. 13; another married the widow of No 12, who was the granddaughter of No. 21. His first wife was a sister of No. 9, and his second wife was a daughter of King John. His stepmother was the aunt of No. 9. His mother's grandfather and the grandfather of No. 6 were brothers. His paternal great-grandfather was a brother of the paternal great-grandfather of No. 6. His paternal great-grandmother was a sister of the maternal grandfather of the wife of No. 6, and the wife of No. 15.

#17. Richard de Montfichet. — His sister was the wife of No. 9. His father's grandfather married a daughter of the first Earl of Clare, ancestor of Nos. 5 and 6. His sister was the mother-in-law of No. 24, and an ancestor married a great-aunt of No. 6.

#18. Roger de Montbegon. — He deserted the Barons soon after being made a Surety (or did he just disappear? wink, wink), and No. 26 was substituted for him. de Montbegon does not seem to have been related by blood or marriage with any of the other Sureties. Possibly feeling a little outnumbered, Roger? And strangely this was already accounted for in the Magna Carta Document in advance of deMontbegon deserting or goes missing... "If-one of the twenty-five barons dies or leaves the country, or is prevented in any other way from discharging his duties, the rest of them shall choose another baron in his place, at their discretion, who shall be duly sworn in as they were." Text from 1215 Magna Carta.

#19. William de Mowbray. — His mother was a sister of the father of No. 6. His wife was a cousin of No. 1. His grandson married a granddaughter of No. 6. His youngest brother, No. 26, was substituted for No. 18.

#20. Richard de Percy. — His mother was a granddaughter of the grandfather of No. 6. His mother was a cousin of No. 6, and also aunt of a son-in-law of No. 4. His step grandmother was also the grandmother to #1 and #22.

#21. Saher de Quincey. — His granddaughter was the second wife of No. 12, and she married, secondly, a brother of No. 16. His son's wife's grandfather was a brother of the maternal grandmother of No. 6. His mother was the widow of a cousin of Nos. 5 and 6, and the paternal grandmother of No. 8. His daughter married a son of No. 24. His wife's aunt was the sister-in-law of the wives of Nos. 6 and 15. His cousin (mother's brother's son, also nephew of the grandmother of No. 8, and cousin of No. 25) married a sister-in-law of the wives of Nos. 6 and 15. His grand- daughter's second husband was a great-grandson of a brother of the grandfather of No. 6, whose wife was a sister of the wife of No. 15.

#22. Robert de Ros. — His grandmother was married to #1's grandfather and to #20's grandfather. His wife was a daughter of King William the Lion whose sister was the wife of No. 25, and a sister-in-law of No. 16. His grandson married a granddaughter of No. I. His multiple times granddaughter married the multiple times grandson of No.4.

#23. Geoffrey de Say. — He was a cousin of No. 15. His grandfather was the grandfather of No. 15. He was related by marriage to No. 9. His paternal grandmother's half-sister was the mother of No. 12. His paternal grandmother was a great-aunt of No. 15

#24. Robert de Vere. — His wife was a niece of No. 17 and was of the family of the first Earl of Clare, ancestor of Nos. 5 and 6. His paternal grandfather's sister's great-aunt — was the first wife of the maternal grandfather of No.12. His son's wife was a daughter of No. 21. His paternal grandmother was a sister of the paternal grandfather of No. 6. His brother married a daughter of No. 3. His father's sister was the mother of No. 3, and the grandmother of No. 2. His paternal grandfather's sister married, secondly, the great-grandfather of No. 15.

#25. Eustace de Vesci. — He was a cousin of No. 7. His wife was also a daughter of the William the Lion who was father to the wife of No. 22, and the sister of the son-in-law of No. 16. His wife's aunt married a cousin of Nos. 8 and 21. His mother-in-law was the great-granddaughter of the brother of the grandfather of No. 6. His wife's sister's husband was a great-grandson of the brother of the grandfather of No. 6.

#26. Roger de Mowbray. — He was substituted for No.18. He was a brother of No. 19, and his mother was an aunt of No. 6. His sister-in-law was a cousin of No. 1.

Magna Carta and Women's Rights in 1215

Chapter 24

Articles 6, 7, 8 of the Magna Carta have to do with women's rights. In 1215 women's rights were written into the Magna Carta. This makes me proud of Robert, taking care of the women folk.

"(6) Heirs may be given in marriage, but not to someone of lower social standing. Before a marriage takes place, it shall be' made known to the heir's next-of-kin."

"(7) At her husband's death, a widow may have her marriage portion and inheritance at once and without trouble. She shall pay nothing for her dower, marriage portion, or any inheritance that she and her husband held jointly on the day of his death. She may remain in her husband's house for forty days after his death, and within this period her dower shall be assigned to her."

"(8) No widow shall be compelled to marry, so long as she wishes to remain without a husband. But she must give security that she will not marry without royal consent, if she holds her lands of the Crown, or without the consent of whatever other lord she may hold them of."

But why are these three articles are included in the 1215 Magna Carta? Throughout European history the land was attached through the women. Meaning if you married a "Landed Woman" you got her land in the Marriage contract. So, if the husband of a Landed Lady died before her, then once again the land continues through her. So, that would make her a prize of sorts to be given out by the King, her father or brothers. Then that being the case, the woman had no choice in the matter of her marriage. So now reread the above 3 articles. The de Ros men starting with Piers de Ros through to Robert II de Ros. Inherited most all their lands from their Aunts or Wives. If the marriage choice were left up to a King then those de Ros lands and Castles would not, could not have been inherited by them. Their lands and the great Castles that the de Ros's were then enjoying. If the above clauses were left out of the Magna Carta, then that would have nullified out their own family inheritances.

Great for women's rights while keeping the Magna Carta Baron's own inherited lands intact. These 3 articles to my eye, protected the Barons own personal estates while taking care of their women as well, a win-win situation of sorts for both, Women's Rights of the day and the 25 Barons of the Magna Carta as well.

His Castle in the Poem Beowulf

Chapter 25

Helmsley's Stern Pile

Photographed by author Susan Rose 2017

"Old force with love will intertwine
In ruin as in strength"

Helmsley HELMSLIE was the name in Angle days of the place which the Normans came to call Hamalac. In the survey of the Temple estates (Temple estates code for Knights Templar's Lands) compiled in 1185 it is written Healmesley. Uncouth as appears the difference between the two words it may be susceptible of explanation as the mere mispronunciation of the Norman whose idea of the sound of the word Lincoln was rendered Nichole.

The place has had an Angle origin as the lega or district of some leader or potentate named Helm. The name of the possessor is not of uncommon application in Yorkshire and between Helmsley and Hemsworth there is very little difference of meaning both of them testifying that the owners of the name had a craving for land and power.

belongs to the Gothic mythology he is said to have ruled the Wylfings who under the name of Helmings Helm's clan or tribesmen are mentioned in Beowulf!

"The Queen of Hrothgar Decked

with gold mindful of kin and name.

She greeted all the men in hall

and to the East Danes.

Lord Joyful she gave the beaker

first and pledged him at the board.

Dear to his folk and blithe of heart

and glad the valiant king.

Partook of feast and banquet cup

the while around the ring of

warriors old and youthful knights

> *the <u>Helmings</u> lady passed.*
> *To each she gave the goblet*
> *rich till by goodhap at last.*
> *The necklaced queen with*
> *courtly grace before Beowulf trod.*
> *Gave him the mead cup*
> *greeted him and offered*
> *thanks to God."*

Of course, Beowulf goes on for pages, the above is only here to reference "The Helmings Lady…" from Helmlac/Hamlake Castle. Robert II's first Castle of his very own… inherited from Walter d'Espic, Robert built his first Castle on this site spoken of in Beowulf.

Names of Robert II de Ros' Castles:

Castle Bonneville-sur-Toques

Castle Helmsley

Castle Warke

Castle Belvoir

Castle Kendal

Plus a Barony in Nairn, Scotland

America's First President, descended from Robert II de Ros

Chapter 26

George Washington, by Gilbert Stuart 1796 Public Domain

Our first president George Washington had a great many ancestors of the landed gentry in the Yorkshire families of England. Of the Magna Carta Sureties, 14 of the Barons are direct ancestors, five are first Cousins, four are Great Uncles, and one Grand Uncle. In all, 24 out of the 25 Sureties he is no less than a fourth cousin.

The 25 Magna Carta Sureties related to George Washington

1. William D' AUBIGNY Lord of Belvoir: Direct Descendant

2. Hugh BIGOD 5th Earl of Norfolk: Direct Descendant

3. Roger BIGOD 4th Earl of Norfolk: Direct Descendant

4. Henry DE BOHUN 5th Earl of Hereford: Direct Descendant

5. Richard DE CLARE 3rd Earl of Hertford: Direct Descendant

6. Gilbert DE CLARE 4th Earl of Hertford: Direct Descendant

7. John FITZ ROBERT Lord of Warkworth: Direct Descendant

8. Robert FITZ WALTER Lord of Dunmow: Direct Descendant

9. William DE FORZ Earl of Aumale: Second Cousin

10. William DE HARDELL, Mayor of London: Ancestry Unknown

11. William DE HUNTINGFIELD Knight: First Cousin

12. John DE LACY 1st Earl of Lincoln: Direct Descendant

13. William III DE LANVALLEI Knight: Fourth Cousin

14. William MALET Lord of Curry Malet: Direct Descendant

15. Geoffrey DE MANDEVILLE 4th Earl of Essex: Great Grand Uncle

16. William MARSHALL 2nd Earl of Pembroke: Great Grand Uncle

17. Roger DE MONTBEGON: Great Grand Uncle

18. Richard DE MONTFICHET Sheriff of Essex: Second Cousin

19. William DE MOWBRAY Baron of Thirsk: Direct Descendant

20. Richard DE PERCY: Great Grand Uncle

21. Saher DE QUINCY 1st Earl of Winchester: Direct Descendant
22. Robert DE ROS Lord of Helmsley: Direct Descendant
23. Geoffrey II DE SAY 2nd Lord of W Greenwich: First Cousin
24. Robert DE VERE 3rd Earl of Oxford: Direct Descendant
25. Eustace DE VESCY Lord of Alnwick: First Cousin

Other Problems of the 3rd Crusade

Chapter 27

By the end of the Battle of Acre during the 3rd Crusade, King Richard was taken down with Scurvy. He was carried around on a litter using a crossbow instead of riding on a horse or using his own 2 feet.

The cause of Scurvy: Vitamin C is a necessary nutrient that helps the body absorb iron and produce collagen. If the body does not produce enough collagen, tissues will start to break down. It is also needed for synthesizing dopamine, norepinephrine, epinephrine, and carnitine, needed for energy production. Symptoms of vitamin C deficiency can start to appear after 8 to 12 weeks. Early signs include a loss of appetite, weight loss, fatigue, irritability, and lethargy. Within 1 to 3 months, there may be signs of:
Anemia, myalgia, or pain, including bone pain, swelling, or edema, petechiae, or small red spots resulting from bleeding under the skin, corkscrew hairs, gum disease and loss of teeth, poor wound healing, shortness of breath, mood changes, and depression

In time, the person will show signs of generalized edema, severe jaundice, destruction of red blood cells, known as hemolysis, sudden and spontaneous bleeding, neuropathy, fever, and convulsions. It can be fatal. There may also be subperiosteal hemorrhage, a type of bleeding that occurs at the ends of the long bones.

Richard's Army won the Battle of Jaffa in 1192. After which Richard rebuilt the fortifications. to the extreme sicknesses of both King Richard and his opponent Saladin, they agreed to talk terms. At which time Richard agreed to demolish the fortifications of Ascalon, and Saladin agreed to allow Crusader control of the coast from Tyre to Jaffa. Christians would have free access to travel as pilgrims to Jerusalem. Saladin died of a fever on 4 March 1193, at Damascus, not long after King Richard's departure.

Saladin had been struck down with Typhoid at the end of the 3rd Crusade, he was not able to participate in the last battles. But by sifting through clues on Saladin's medical symptoms written more than 800 years ago, a doctor may have finally determined what illness felled the mighty sultan.

It was typhoid, said Dr. Stephen Gluckman, a professor of medicine at the University of Pennsylvania's Perelman School of Medicine, announced today (May 4) at the 25th annual Historical Clinicopathological Conference at the University of Maryland School of Medicine.

Typhoid is a bacterial infection that can lead to a high fever, diarrhea, and vomiting. It can be fatal. It is caused by the bacteria Salmonella typhi. The infection is often passed on through contaminated food and drinking water, and it is more prevalent in places where handwashing is less frequent. It can alsTyphoid is an infection caused by the bacterium Salmonella typhimurium (S. typhi).

The bacterium lives in the intestines and bloodstream of humans. It spreads between individuals by direct contact with the feces of an infected person. No animals carry this disease, so transmission is always human to human.

If untreated, around 1 in 5 cases of typhoid can be fatal. With treatment, fewer than 4 in 100 cases are fatal. S. typhi enters through the mouth and spends 1 to 3 weeks in the intestine. After this, it makes its way through the intestinal wall and into the bloodstream. From the bloodstream, it spreads into other tissues and organs. The immune system of the host can do little to fight back because S. typhi can live within the host's cells, safe from the immune system.

Patron Saints of the LGBTQ
Chapter 28

One cannot miss the accusations by the Catholic Pope during the Knights Templar Trials 1308, a reference to homosexual activities by the Knights Templar. Rievaulx Abbey and nearby Ribstan Abbeys were Knights Templar holdings at that time. It is interesting that the Knights Templar coordinating architect was Bernard de Clairvaux, long thought to have been gay and the head Abbot Aelred of Rievaulx is also considered to have been gay. Both are Patron Saints to the current day LGBTQ community. Yet the Knights Templar organization of that day looks to not have had a problem what-so-ever by these men. What counted to the Templars was the quality of the men themselves. It seems that the Knights Templar were before their times.

Patron Saints of the current LGBTQ Community are Saint Aelred of Rievaulx and Saint Bernard de Clairvaux.

S. AELRED

Saint Aelred of Rievaulx, Front Piece to Lives of the English Saints, by John Henry Newman 1845 Public Doman

Saint Aelred of Rievaulx This 12-century abbot served in the court of Kind David of Scotland until 1134, when he entered the Cistercian abbey of Rievaulx. He went on to be the Abbot of Rievaulx. Known as an Historian rather than his Religious works. He was confirmed a Saint prior to The rules of Canonization that were organized by Gregory IX 1227-1241. He is considered a Saint who died prior to the Pentecost and thusly is a saint of Old Covenant.

Saint Bernard de Clairvaux Public Domain

Saint Bernard of Clairvaux an abbot in medieval France, Bernard of Clairvaux maintained a lengthy personal relationship with the Archbishop Malachy. After Malachy, an Irishman who also became a saint, died in Bernard's arms, Bernard wore the fallen religious leader's habits for his remaining Years. He was the primary force in founding of the Cistercian Order of Monks, for the founding of Rievaulx Abbey and writing the Rules of the Templar Knights. He became canonized by Pope Alexander III on January 18, 1174. Bernard of Clairvaux, from

Project Gutenburg from "a Short History of Monks and Monasteries by Alfred Wesley Wishart

Peacocks and Knights Templar

Chapter 29

Carved Peacock, at Rievaulx Abbey photographed by author Susan Rose 2017

The Knights Templar Imagery of the Peacock. Imagery used to notate Knights Templar sites around England and Ireland is the Peacock on gravestones, heraldry, religious houses and castles. Gives clues to the casual observer that it is a Knights Templar site.

A stained-glass window by C E Kempe & Co. depicting Saint Michael Archangel slaying the devil. Alamy

Non-Christian religions held the belief that the peacock's flesh never decayed, even after it died. Early Christians, therefore, adopted the bird as a symbol of the Resurrection, Christ's eternal, glorious existence. In medieval times, it was also thought that peacocks molt (shed their feathers) every year, and the new ones that grow are more beautiful than the old ones.

The "eyes" in the peacock's tail feathers symbolize the all-seeing God, as the peacock feather has eyes both on the front of the feather and the back.

Peacock sculptures were located around Rievaulx Abbey and can still be seen on the inside and outside of Belvoir Castle today.

Rules of the Knights Templar

Chapter 30

The Primitive Rule of the Templars:

Trans. Mrs. Judith Upton-Ward (Reprinted by kind permission of the author) This translation of the original, or primitive, Rule of the Templars is based on the 1886 edition of Henri de Curzon, La Régle du Temple as a Military Manual, or How to Deliver a Cavalry Charge.

It represents the Rule given to the fledgling Knights of the Temple by the Council of Troyes, 1129, although "it must not be forgotten that the Order had been in existence for several years and had built up its own traditions and customs before Hugues de Payens' appearance at the Council of Troyes.

To a considerable extent, then, the Primitive Rule is based upon existing practices." (Upton-Ward, p. 11) This translation is excerpted from Judith Upton-Ward's The Rule of the Templars, Woodbridge: The Boydell Press, 1992, and is reprinted here with permission. The Rule of the Templars includes an introduction by Upton-Ward; it also contains the Templars' Primitive Rule and the Hierarchical Statutes; regulations governing penances, conventual life, the holding of ordinary chapters, and reception into the Order; and an appendix by Matthew Bennett, "La Régle du Temple as a Military Manual, or How to Deliver a Cavalry Charge." The book is highly recommended to those interested in the Templars or any other military order. It is now available in paperback. The notes to the Primitive Rule, supplied by Mrs. Upton-Ward in The Rule of the Templars, are not included below. They are of considerable interest and should be consulted by those wishing to study the Rule in more detail, however.

The Primitive Rule of the Templars THE PRIMITIVE RULE Here begins the prologue to the Rule of Temple

1. We speak firstly to all those who secretly despise their own will and desire with a pure heart to serve the sovereign king as a knight and with studious care desire to wear, and wear permanently, the very noble armour of obedience. And therefore we admonish you, you who until now have led the lives of secular knights, in which Jesus Christ was not the cause, but which you embraced for human favour only, to follow those whom God has chosen from the mass of perdition and whom he has ordered through his gracious mercy to defend the Holy Church, and that you hasten to join them forever.

2. Above all things, whosoever would be a knight of Christ, choosing such holy orders, you in your profession of faith must unite pure diligence and firm perseverence, which is so worthy and so holy, and is known to be so noble, that if it is preserved untainted for ever, you will deserve to keep company with the martyrs who gave their souls for Jesus Christ. In this religious order has flourished and is revitalised the order of knighthood. This knighthood despised the love of justice that constitutes its duties and did not do what it should, that is defend the poor, widows, orphans and churches, but strove to plunder, despoil and kill. God works well with us and our saviour Jesus Christ; He has sent his friends from the Holy City of Jerusalem to the marches of France and Burgundy, who for our salvation and the spread of the true faith do not cease to offer their souls to God, a welcome sacrifice.

3. Then we, in all joy and all brotherhood, at the request of Master Hugues de Payens, by whom the aforementioned knighthood was founded by the grace of the Holy Spirit, assembled at Troyes from divers provinces beyond the mountains on the feast of my lord St Hilary, in the year of the incarnation of Jesus Christ 1128, in the ninth year after the founding of the aforesaid knighthood. And the conduct and beginnings of the Order of Knighthood we heard in common chapter from the lips of the aforementioned Master, Brother Hugues de Payens; and according to the limitations of our understanding what seemed to us good and beneficial we praised, and what seemed wrong we eschewed.

4. And all that took place at that council cannot be told nor recounted; and so that it should not be taken lightly by us, but considered in wise prudence, we left it to the discretion of both our honourable father lord Honorius and of the noble patriarch of Jerusalem, Stephen, who knew the affairs of the East and of the Poor Knights of Christ, by the advice of the common council we praised it unanimously. Although a great number of religious fathers who assembled at that council praised the authority of our words, nevertheless we should not pass over in silence the true sentences and judgements which they pronounced.

5. Therefore I, Jean Michel, to whom was entrusted and confided that divine office, by the grace of God served as the humble scribe of the present document by order of the council and of the venerable father Bernard, abbot of Clairvaux.

The Names of the Fathers who Attended the Council

6. First was Matthew, bishop of Albano, by the grace of God legate of the Holy Church of Rome; R[enaud], archbishop of Reims; H(enri), archbishop of Sens; and then their suffragans: G(ocelin], bishop of Soissons; the bishop of Paris; the bishop of Troyes; the bishop of Orlèans; the bishop of Auxerre; the bishop of Meaux; the bishop of Chalons; the bishop of Laon; the bishop of Beauvais; the abbot of Vèzelay, who was later made archbishop of Lyon and legate of the Church of Rome; the abbot of Cîteaux; the abbot of Pontigny; the abbot of Trois-Fontaines; the abbot of St Denis de Reims; the abbot of St-Etienne de Dijon; the abbot of Molesmes; the above-named B[ernard], abbot of Clairvaux: whose words the aforementioned praised liberally. Also present were master Aubri de Reims; master Fulcher and several others whom it would be tedious to record. And of the others who have not been listed it seems profitable to furnish guarantees in this matter, that they are lovers of truth: they are count Theobald; the count of Nevers; Andrè de Baudemant. These were at the council and acted in such a manner that by perfect, studious care they sought out that which was fine and disapproved that which did not seem right.

7. And also present was Brother Hugues de Payens, Master of the Knighthood, with some of his brothers whom he had brought with him. They were Brother Roland, Brother Godefroy, and Brother Geoffroi Bisot, Brother Payen de Montdidier, Brother Archambaut de Saint-Amand. The same Master Hugues with his followers related to the above-named fathers the customs and observances of their humble beginnings and of the one who said: Ego principium qui et loquor vobis, that is to say: 'I who speak to you am the beginning,' according to one's memory.

8. It pleased the common council that the deliberations which were made there and the consideration of the Holy Scriptures which were diligently examined with the wisdom of my lord H[onorius], pope of the Holy Church of Rome, and of the patriarch of Jerusalem and with the assent of the chapter, together with the agreement of the Poor Knights of Christ of the Temple which is in Jerusalem, should be put in writing and not forgotten, steadfastly kept so that by an upright life one may come to his creator; the compassion of which Lord [is sweeter] than honey when compared with God; whose mercy resembles oine, and permits us to come to Him whom they desire to serve. Per infinita seculorum secula. Amen

Here Begins the Rule of the Poor Knighthood of the Temple

9. You who renounce your own wills, and you others serving the sovereign king with horses and arms, for the salvation of your souls, for a fixed term, strive everywhere with pure desire to hear matins and the entire service according to canonical law and the customs of the regular masters of the Holy City of Jerusalem. 0 you venerable brothers, similarly God is with you, if you promise to despise the deceitful world in perpetual love of God, and scorn the temptations of your body: sustained by the food of God and watered and instructed in the commandments of Our Lord, at the end of the divine office, none should fear to go into battle if he henceforth wears the tonsure.

10. But if any brother is sent through the work of the house and of Christianity in the East--something we believe will happen often--and cannot hear the divine office, he should say instead of matins thirteen

paternosters; seven for each hour and nine for vespers. And together we all order him to do so. But those who are sent for such a reason and cannot come at the hours set to hear the divine office, if possible the set hours should not be omitted, in order to render to God his due.

The Manner in which Brothers should be Received

11. If any secular knight, or any other man, wishes to leave the mass of perdition and abandon that secular life and choose your communal life, do not consent to receive him immediately, for thus said my lord St Paul: Probate spiritus si ex Deo sunt. That is to say: 'Test the soul to see if it comes from God.' Rather, if the company of the brothers is to be granted to him, let the Rule be read to him, and if he wishes to studiously obey the commandments of the Rule, and if it pleases the Master and the brothers to receive him, let him reveal his wish and desire before all the brothers assembled in chapter and let him make his request with a pure heart.

On Excommunicated Knights

12. Where you know excommunicated knights to be gathered, there we command you to go; and if anyone there wishes to join the order of knighthood from regions overseas, you should not consider worldly gain so much as the eternal salvation of his soul. We order him to be received on condition that he come before the bishop of that province and make his intention known to him. And when the bishop has heard and absolved him, he should send him to the Master and brothers of the Temple, and if his life is honest and worthy of their company, if he seems good to the Master and brothers, let him be mercifully received; and if he should die in the meanwhile, through the anguish and torment he has suffered, let him be given all the benefits of the brotherhood due to one of the Poor Knights of the Temple.

13. Under no other circumstances should the brothers of the Temple share the company of an obviously-excommunicated man, nor take his own things; and this we prohibit strongly because it would be a fearful thing if they were excommunicated like him. But if he is only forbidden to hear the divine office, it is certainly possible to keep company with

him and take his property for charity with the permission of their commander.

On Not Receiving Children

14. Although the rule of the holy fathers allows the receiving of children into a religious life, we do not advise you to do this. For he who wishes to give his child eternally to the order of knighthood should bring him up until such time as he is able to bear arms with vigour, and rid the land of the enemies of Jesus Christ. Then let the mother and father lead him to the house and make his request known to the brothers; and it is much better if he does not take the vow when he is a child, but when he is older, and it is better if he does not regret it than if he regrets it. And henceforth let him be put to the test according to the wisdom of the Master and brothers and according to the honesty of the life of the one who asks to be admitted to the brotherhood.

On Brothers who Stand Too Long in Chapel

15. It has been made known to us and we heard it from true witnesses that immoderately and without restraint you hear the divine service whilst standing. We do not ordain that you behave in this manner, on the contrary we disapprove of it. But we command that the strong as well as the weak, to avoid a fuss, should sing the psalm which is called Venite, with the invitatory and the hymn sitting down, and say their prayers in silence, softly and not loudly, so that the proclaimer does not disturb the prayers of the other brothers.

16. But at the end of the psalms, when the Gloria patri is sung, through reverence for the Holy Trinity, you will rise and bow towards the altar, while the weak and ill will incline their heads. So we command; and when the explanation of the Gospels is read, and the Te deum laudamus is sung, and while all the lauds are sung, and the matins are finished, you will be on your feet. In such a manner we command you likewise to be on your feet at matins and at all the hours of Our Lady.

On the Brothers' Dress 17. We command that all the brothers' habits should always be of one colour, that is white or black or brown. And we grant to all knight brothers in winter and in summer if possible, white cloaks; and no-one who does not belong to the aforementioned Knights of Christ is allowed to have a white cloak, so that those who have abandoned the life of darkness will recognise each other as being reconciled to their creator by the sign of the white habits: which signifies purity and complete chastity. Chastity is certitude of heart and healthiness of body. For if any brother does not take the vow of chastity he cannot come to eternal rest nor see God, by the promise of the apostle who said: Pacem sectamini cum omnibus et castimoniam sine qua nemo Deum videbit. That is to say: 'Strive to bring peace to all, keep chaste, without which no-one can see God.'

18. But these robes should be without any finery and without any show of pride. And so we ordain that no brother will have a piece of fur on his clothes, nor anything else which belongs to the usages of the body, not even a blanket unless it is of lamb's wool or sheep's wool. We command all to have the same, so that each can dress and undress, and put on and take off his boots easily. And the Draper or the one who is in his place should studiously reflect and take care to have the reward of God in all the above-mentioned things, so that the eyes of the envious and evil-tongued cannot observe that the robes are too long or too short; but he should distribute them so that they fit those who must wear them, according to the size of each one.

19. And if any brother out of a feeling of pride or arrogance wishes to have as his due a better and finer habit, let him be given the worst. And those who receive new robes must immediately return the old ones, to be given to the squires and sergeants and often to the poor, according to what seems good to the one who holds that office.

On Shirts

20. Among the other things, we mercifully rule that, because of the great intensity of the heat which exists in the East, from Easter to All Saints, through compassion and in no way as a right, a linen shirt shalt be given to any brother who wishes to wear it.

On Bed Linen

21. We command by common consent that each man shall have clothes and bed linen according to the discretion of the Master. It is our intention that apart from a mattress, one bolster and one blanket should be sufficient for each; and he who lacks one of these may have a rug, and he may use a linen blanket at all times, that is to say with a soft pile. And they will at all times sleep dressed in shirt and breeches and shoes and belts, and where they sleep shall be lit until morning. And the Draper should ensure that the brothers are so well tonsured that they may be examined from the front and from behind; and we command you to firmly adhere to this same conduct with respect to beards and moustaches, so that no excess may be noted on their bodies.

On Pointed Shoes' and Shoe-Laces

22. We prohibit pointed shoes and shoe-laces and forbid any brother to wear them; nor do we permit them to those who serve the house for a fixed term; rather we forbid them to have shoes with points or laces under any circumstances. For it is manifest and well known that these abominable things belong to pagans. Nor should they wear their hair or their habits too long. For those who serve the sovereign creator must of necessity be born within and without through the promise of God himself who said: Estote mundi quia ego mundus sum. That is to say: 'Be born as I am born.'

How They Should Eat

23. In the palace, or what should rather be called the refectory, they should eat together. But if you are in need of anything because you are not accustomed to the signs used by other men of religion, quietly and privately you should ask for what you need at table, with all humility and submission. For the apostle said: Manduca panem tuum cum silentio. That is to say: 'Eat your bread in silence.' And the psalmist: Posui ori meo custodiam. That is to say: 'I held my tongue.' That is, 'I thought my tongue would fail me.' That is, 'I held my tongue so that I should speak no ill.'

On the Reading of the Lesson

24. Always, at the convent's dinner and supper, let the Holy Scripture be read, if possible. If we love God and all His holy words and His holy commandments, we should desire to listen attentively; the reader of the lesson will tell you to keep silent before he begins to read.

On Bowls and Drinking Vessels

25. Because of the shortage of bowls, the brothers will eat in pairs, so that one may study the other more closely, and so that neither austerity nor secret abstinence is introduced into the communal meal. And it seems just to us that each brother should have the same ration of wine in his cup. On the Eating of Meat 26. It should be sufficient for you to eat meat three times a week, except at Christmas, All Saints, the Assumption and the feast of the twelve apostles. For it is understood that the custom of eating flesh corrupts the body. But if a fast when meat must be forgone falls on a Tuesday, the next day let it be given to the brothers in plenty. And on Sundays all the brothers of the Temple, the chaplains and the clerks shall be given two meat meals in honour of the holy resurrection of Jesus Christ. And the rest of the household, that is to say the squires and sergeants, shall be content with one meal and shall be thankful to God for it.

On Weekday Meals

27. On the other days of the week, that is Mondays, Wednesdays and even Saturdays, the brothers shall have two or three meals of vegetables or other dishes eaten with bread; and we intend that this should be sufficient and command that it should be adhered to. For he who does not eat one meal shall eat the other.

On Friday Meals

28. On Fridays, let lenten meat be given communally to the whole congregation, out of reverence for the passion of Jesus Christ; and you will fast from All Saints until Easter, except for Christmas Day, the Assumption and the feast of the twelve apostles. But weak and sick brothers shall not be kept to this. From Easter to All Saints they may eat twice, as long as there is no general fast.

On Saying Grace

29. Always after every dinner and supper all the brothers should give thanks to God in silence, if the church is near to the palace where they eat, and if it is not nearby, in the place itself. With a humble heart they should give thanks to Jesus Christ who is the Lord Provider. Let the remains of the broken bread be given to the poor and whole loaves be kept. Although the reward of the poor, which is the kingdom of heaven, should be given to the poor without hesitation, and the Christian faith doubtless recognises you among them, we ordain that a tenth part of the bread be given to your Almoner.

On Taking Collation

30. When daylight fades and night falls listen to the signal of the bell or the call to prayers, according to the customs of the country, and all go to compline. But we command you first to take collation; although we place this light meal under the arbitration and discretion of the Master. When he wants water and when he orders, out of mercy, diluted wine, let it be given sensibly. Truly, it should not be taken to excess, but in moderation. For Solomon said: Quia vinum facit apostatare sapientes.ÃÃ ÄÄThat is to say that wine corrupts the wise.

On Keeping Silence

31. When the brothers come out of compline they have no permission to speak openly except in an emergency. But let each go to his bed quietly and in silence, and if he needs to speak to his squire, he should say what he has to say softly and quietly. But if by chance, as they come out of compline, the knighthood or the house has a serious problem which must be solved before morning, we intend that the Master or a party of elder brothers who govern the Order under the Master, may speak appropriately. And for this reason we command that it should be done in such a manner.

32. For it is written: In multiloquio non effugies peccatum. That is to say that to talk too much is not without sin. And elsewhere: Mors et vita in manibus lingue. That is to say: 'Life and death are in the power of the tongue.' And during that conversation we altogether prohibit idle

words and wicked bursts of laughter. And if anything is said during that conversation that should not be said, when you go to bed we command you to say the paternoster prayer in all humility and pure devotion.

On Ailing Brothers

33. Brothers who suffer illness through the work of the house may be allowed to rise at matins with the agreement and permission of the Master or of those who are charged with that office. But they should say instead of matins thirteen paternosters, as is established above, in such a manner that the words reflect the heart. Thus said David: Psallite sapienter. That is to say: 'Sing wisely.' And elsewhere the same David said: In conspectu Angelorum psallam tibi. That is to say: 'I will sing to you before the angels.' And let this thing be at all times at the discretion of the Master or of those who are charged with that office.

On the Communal Life

34. One reads in the Holy Scriptures: Dividebatur singulis prout cuique opus erat. That is to say that to each was given according to his need. For this reason we say that no-one should be elevated among you, but all should take care of the sick; and he who is less ill should thank God and not be troubled; and let whoever is worse humble himself through his infirmity and not become proud through pity. In this way all members will live in peace. And we forbid anyone to embrace excessive abstinence; but firmly keep the communal life. On the Master

35. The Master may give to whomsoever he pleases the horse and armour and whatever he likes of another brother, and the brother to whom the given thing belongs should not become vexed or angry: for be certain that if he becomes angry he will go against God. On Giving Counsel

36. Let only those brothers whom the Master knows will give wise and beneficial advice be called to the council; for this we command, and by no means everyone should be chosen. For when it happens that they wish to treat serious matters like the giving of communal land, or to speak of the affairs of the house, or receive a brother, then if the Master wishes, it is appropriate to assemble the entire congregation to

hear the advice of the whole chapter; and what seems to the Master best and most beneficial, let him do it.

On Brothers Sent Overseas

37. Brothers who are sent throughout divers countries of the world should endeavour to keep the commandments of the Rule according to their ability and live without reproach with regard to meat and wine, etc. so that they may receive a good report from outsiders and not sully by deed or word the precepts of the Order, and so that they may set an example of good works and wisdom; above all so that those with whom they associate and those in whose inns they lodge may be bestowed with honour. And if possible, the house where they sleep and take lodging should not be without light at night, so that shadowy enemies may not lead them to wickedness, which God forbids them.

On Keeping the Peace

38. Each brother should ensure that he does not incite another brother to wrath or anger, for the sovereign mercy of God holds the strong and weak brother equal, in the name of charity. How the Brothers Should Go About

39. In order to carry out their holy duties and gain the glory of the Lord's joy and to escape the fear of hell-fire, it is fitting that all brothers who are professed strictly obey their Master. For nothing is dearer to Jesus Christ than obedience. For as soon as something is commanded by the Master or by him to whom the Master has given the authority, it should be done without delay as though Christ himself had commanded it. For thus said Jesus Christ through the mouth of David, and it is true: Ob auditu auris obedivit mihi. That is to say: 'He obeyed me as soon as he heard me.'

40. For this reason we pray and firmly command the knight brothers who have abandoned their own wills and all the others who serve for a fixed term not to presume to go out into the town or city without the permission of the Master or of the one who is given that office; except at night to the Sepulchre and the places of prayer which lie within the walls of the city of Jerusalem.

41. There, brothers may go in pairs, but otherwise may not go out by day or night; and when they have stopped at an inn, neither brother nor squire nor sergeant may go to another's lodging to see or speak to him without permission, as is said above. We command by common consent that in this Order which is ruled by God, no brother should fight or rest according to his own will, but according to the orders of the Master, to whom all should submit, that they may follow this pronouncement of Jesus Christ who said: Non veni facere voluntatem meam, sed ejus que misit me, patris. That is to say: 'I did not come to do my own will, but the will of my father who sent me.'

How they should Effect an Exchange

42. Without permission from the Master or from the one who holds that office, let no brother exchange one thing for another, nor ask to, unless it is a small or petty thing. On Locks

43. Without permission from the Master or from the one who holds that office, let no brother have a lockable purse or bag; but commanders of houses or provinces and Masters shall not be held to this. Without the consent of the Master or of his commander, let no brother have letters from his relatives or any other person; but if he has permission, and if it please the Master or the commander, the letters may be read to him.

On Secular Gifts

44. If anything which cannot be conserved, like meat, is given to any brother by a secular person in thanks, he should present it to the Master or the Commander of Victuals. But if it happens that any of his friends or relatives has something that they wish to give only to him, let him not take it without the permission of the Master or of the one who holds that office. Moreover, if the brother is sent any other thing by his relatives, let him not take it without the permission of the Master or of the one who holds that office. We do not wish the commanders or baillis, who are especially charged to carry out this office, to be held to this aforementioned rule.

On Faults

45. If any brother, in speaking or soldiering, or in any other way commits a slight sin, he himself should willingly make known the fault to the Master, to make amends with a pure heart. And if he does not usually fail in this way let him be given a light penance, but if the fault is very serious let him go apart from the company of the brothers so that he does not eat or drink at any table with them, but all alone; and he should submit to the mercy and judgement of the Master and brothers, that he may be saved on the Day of Judgement. On Serious Faults

46. Above all things, we should ensure that no brother, powerful or not powerful, strong or weak, who wishes to promote himself gradually and become proud and defend his crime, remain unpunished. But if he does not wish to atone for it let him be given a harsher punishment. And if by pious counsel prayers are said to God for him, and he does not wish to make amends, but wishes to boast more and more of it, let him be uprooted from the pious flock; according to the apostle who says: Auferte malum ex vobis. That is to say: 'Remove the wicked from among you.' It is necessary for you to remove the wicked sheep from the company of faithful brothers. 47. Moreover the Master, who should hold in his hand the staff and rod- the staff with which to sustain the weaknesses and strengths of others; the rod with which to beat the vices of those who sin--for love of justice by counsel of the patriarch, should take care to do this. But also, as my lord St Maxime said: 'May the leniency be no greater than the fault; nor excessive punishment cause the sinner to return to evil deeds.'

On Rumour

48. We command you by divine counsel to avoid a plague: envy, rumour, spite, slander. So each one should zealously guard against what the apostle said: Ne sis criminator et susurro in populo. That is to say: 'Do not accuse or malign the people of God.' But when a brother knows for certain that his fellow brother has sinned, quietly and with fraternal mercy let him be chastised privately between the two of them, and if he does not wish to listen, another brother should be called, and if he scorns them both he should recant openly before the whole

chapter. Those who disparage others suffer from a terrible blindness and many are full of great sorrow that they do not guard against harbouring envy towards others; by which they shall be plunged into the ancient wickedness of the devil.

Let None Take Pride in his Faults

49. Although all idle words are generally known to be sinful, they will be spoken by those who take pride in their own sin before the strict judge Jesus Christ; which is demonstrated by what David said: Obmutui et silui a bonis. That is to say that one should refrain from speaking even good, and observe silence. Likewise one should guard against speaking evil, in order to escape the penalty of sin. We prohibit and firmly forbid any brother to recount to another brother nor to anyone else the brave deeds he has done in secular life, which should rather be called follies committed in the performance of knightly duties, and the pleasures of the flesh that he has had with immoral women; and if it happens that he hears them being told by another brother, he should immediately silence him; and if he cannot do this, he should straightaway leave that place and not give his heart's ear to the pedlar of filth. Let None Ask

50. This custom among the others we command you to adhere to strictly and firmly: that no brother should explicitly ask for the horse or armour of another. It will therefore be done in this manner: if the infirmity of the brother or the frailty of his animals or his armour is known to be such that the brother cannot go out to do the work of the house without harm, let him go to the Master, or to the one who is in his place in that office after the Master, and make the situation known to him in pure faith and true fraternity, and henceforth remain at the disposal of the Master or of the one who holds that office. On Animals and Squires

51. Each knight brother may have three horses and no more without the permission of the Master, because of the great poverty which exists at the present time in the house of God and of the Temple of Solomon. To each knight brother we grant three horses and one squire, and if that squire willingly serves charity, the brother should not beat

him for any sin he commits. That No Brother May Have an Ornate Bridle

52. We utterly forbid any brother to have gold or silver on his bridle, nor on his stirrups, nor on his spurs. That is, if he buys them; but if it happens that a harness is given to him in charity which is so old that the gold or silver is tarnished, that the resplendent beauty is not seen by others nor pride taken in them: then he may have them. But if he is given new equipment let the Master deal with it as he sees fit. On Lance Covers

53. Let no brother have a cover on his shield or his lance, for it is no advantage, on the contrary we understand that it would be very harmful.

On Food Bags

54. This command which is established by us it is beneficial for all to keep and for this reason we ordain that it be kept henceforth, and that no brother may make a food bag of linen or wool, principally, or anything else except a profinel.

On Hunting

55. We collectively forbid any brother to hunt a bird with another bird. It is not fitting for a man of religion to succumb to pleasures, but to hear willingly the commandments of God, to be often at prayer and each day to confess tearfully to God in his prayers the sins he has committed. No brother may presume to go particularly with a man who hunts one bird with another. Rather it is fitting for every religious man to go simply and humbly without laughing or talking too much, but reasonably and without raising his voice and for this reason we command especially all brothers not to go in the woods with longbow or crossbow to hunt animals or to accompany anyone who would do so, except out of love to save him from faithless pagans. Nor should you go after dogs, nor shout or chatter, nor spur on a horse out of a desire to capture a wild beast. On the Lion

56. It is the truth that you especially are charged with the duty of giving your souls for your brothers, as did Jesus Christ, and of

defending the land from the unbelieving pagans who are the enemies of the son of the Virgin Mary. This above-mentioned prohibition of hunting is by no means intended to include the lion, for he comes encircling and searching for what he can devour, his hands against every man and every man's hand against him. How They May Have Lands and Men 57. This kind of new order we believe was born out of the Holy Scriptures and divine providence in the Holy Land of the East. That is to say that this armed company of knights may kill the enemies of the cross without sinning. For this reason we judge you to be rightly called knights of the Temple, with the double merit and beauty of probity, and that you may have lands and keep men, villeins and fields and govern them justly, and take your right to them as it is specifically established. On Tithes

58. You who have abandoned the pleasant riches of this world, we believe you to have willingly subjected yourselves to poverty; therefore we are resolved that you who live the communal life may receive tithes. If the bishop of the place, to whom the tithe should be rendered by right, wishes to give it to you out of charity, with the consent of his chapter he may give those tithes which the Church possesses. Moreover, if any layman keeps the tithes of his patrimony, to his detriment and against the Church, and wishes to leave them to you, he may do so with the permission of the prelate and his chapter.

On Giving Judgement

59. We know, because we have seen it, that persecutors and people who like quarrels and endeavour to cruelly torment those faithful to the Holy Church and their friends, are without number. By the clear judgement of our council, we command that if there is anyone in the parties of the East or anywhere else who asks anything of you, for faithful men and love of truth you should judge the thing, if the other party wishes to allow it. This same commandment should be kept at all times when something is stolen from you. Elderly Brothers

60. We command by pious counsel that ageing and weak brothers be honoured with diligence and given consideration according to their

frailty; and, kept well by the authority of the Rule in those things which are necessary to their physical welfare, should in no way be in distress.

On Sick Brothers

61. Let sick brothers be given consideration and care and be served according to the saying of the evangelist and Jesus Christ: Infirmus fui et visitastis me. That is to say: 'I was sick and you visited me'; and let this not be forgotten. For those brothers who are wretched should be treated quietly and with care, for which service, carried out without hesitation, you will gain the kingdom of heaven. Therefore we command the Infirmarer to studiously and faithfully provide those things which are necessary to the various sick brothers, such as meat, flesh, birds and all other foods which bring good health, according to the means and the ability of the house.

On Deceased Brothers

62. When any brother passes from life to death, a thing from which no one is exempt, we command you to sing mass for his soul with a pure heart, and have the divine office performed by the priests who serve the sovereign king and you who serve charity for a fixed term and all the brothers who are present where the body lies and serve for a fixed term should say one hundred paternosters during the next seven days. And all the brothers who are under the command of that house where the brother has passed away should say the hundred paternosters, as is said above, afrer the death of the brother is known, by God's mercy. Also we pray and command by pastoral authority that a pauper be fed with meat and wine for forty days in memory of the dead brother, just as if he were alive. We expressly forbid all other offerings which used to be made at will and without discretion by the Poor Knights of the Temple on the death of brothers, at the feast of Easter and at other feasts.

63. Moreover, you should profess your faith with a pure heart night and day that you may be compared in this respect to the wisest of all the prophets, who said: Calicem salutaris accipiam. That is to say: 'I will take the cup of salvation.' Which means: 'I will avenge the death of

Jesus Christ by my death. For just as Jesus Christ gave his body for me, I am prepared in the same way to give my soul for my brothers.' This is a suitable offering; a living sacrifice and very pleasing to God. On the Priests and Clerks who Serve Charity

64. The whole of the common council commands you to render all offerings and all kinds of alms in whatever manner they may be given, to the chaplains and clerks and to others who remain in charity for a fixed term. According to the authority of the Lord God, the servants of the Church may have only food and clothing, and may not presume to have anything else unless the Master wishes to give them anything willingly out of charity.

On Secular Knights 65. Those who serve out of pity and remain with you for a fixed term are knights of the house of God and of the Temple of Solomon; therefore out of pity we pray and finally command that if during his stay the power of God takes any one of them, for love of God and out of brotherly mercy, one pauper be fed for seven days for the sake of his soul, and each brother in that house should say thirty paternosters.

On Secular Knights who Serve for a Fixed Term

66. We command all secular knights who desire with a pure heart to serve Jesus Christ and the house of the Temple of Solomon for a fixed term to faithfully buy a suitable horse and arms, and everything that will be necessary for such work. Furthermore, we command both parties to put a price on the horse and to put the price in writing so that it is not forgotten; and let everything that the knight, his squire and horse need, even horseshoes, be given out of fraternal charity according to the means of the house. If, during the fixed term, it happens by chance that the horse dies in the service of the house, if the house can afford to, the Master should replace it. If, at the end of his tenure, the knight wishes to return to his own country, he should leave to the house, out of charity, half the price of the horse, and the other half he may, if he wishes, receive from the alms of the house.

On the Commitment of Sergeants

67. As the squires and sergeants who wish to serve charity in the house of the Temple for the salvation of their souls and for a fixed term come from divers regions, it seems to us beneficial that their promises be received, so that the envious enemy does not put it in their hearts to repent of or renounce their good intentions. On White Mantles

68. By common counsel of all the chapter we forbid and order expulsion, for common vice, of anyone who without discretion was in the house of God and of the Knights of the Temple; also that the sergeants and squires should not have white habits, from which custom great harm used to come to the house; for in the regions beyond the mountains false brothers, married men and others who said they were brothers of the Temple used to be sworn in; while they were of the world. They brought so much shame to us and harm to the Order of Knighthood that even their squires boasted of it; for this reason numerous scandals arose. Therefore let them assiduously be given black robes; but if these cannot be found, they should be given what is available in that province; or what is the least expensive, that is burell.

On Married Brothers

69. If married men ask to be admitted to the fraternity, benefice and devotions of the house, we permit you to receive them on the following conditions: that after their death they leave you a part of their estate and all that they have obtained henceforth. Meanwhile, they should lead honest lives and endeavour to act well towards the brothers. But they should not wear white habits or cloaks; moreover, if the lord should die before his lady, the brothers should take part of his estate and let the lady have the rest to support her during her lifetime; for it does not seem right to us that such confréres should live in a house with brothers who have promised chastity to God.

On Sisters 70. The company of women is a dangerous thing, for by it the old devil has led many from the straight path to Paradise. Henceforth, let not ladies be admitted as sisters into the house of the Temple; that is why, very dear brothers, henceforth it is not fitting to follow this custom, that the flower of chastity is always maintained among you.

Let Them Not Have Familiarity with Women

71. We believe it to be a dangerous thing for any religious to look too much upon the face of woman. For this reason none of you may presume to kiss a woman, be it widow, young girl, mother, sister, aunt or any other; and henceforth the Knighthood of Jesus Christ should avoid at all costs the embraces of women, by which men have perished many times, so that they may remain eternally before the face of God with a pure conscience and sure life. Not Being Godfathers

72. We forbid all brothers henceforth to dare to raise children over the font and none should be ashamed to refuse to be godfathers or godmothers; this shame brings more glory than sin.

On the Commandments

73. All the commandments which are mentioned and written above in this present Rule are at the discretion and judgement of the Master.

These are the Feast Days and Fasts that all the Brothers should Celebrate and Observe

74. Let it be known to all present and future brothers of the Temple that they should fast at the vigils of the twelve apostles. That is to say: St Peter and St Paul; St Andrew; St James and St Philip; St Thomas; St Bartholomew; Sts. Simon and Jude St James; St Matthew. The vigil of St John the Baptist; the vigil of the Ascension and the two days before, the rogation days; the vigil of Pentecost; the ember days; the vigil of St Laurence; the vigil of Our Lady in mid-August; the vigil of All Saints; the vigil of Epiphany. And they should fast on all the above-mentioned days according to the commandments of Pope Innocent at the council which took place in the city of Pisa. And if any of the above-mentioned feast days fall on a Monday, they should fast on the preceding Saturday. If the nativity of Our Lord falls on a Friday, the brothers should eat meat in honour of the festival. But they should fast on the feast day of St Mark because of the Litany: for it is established by Rome for the mortality of men. However, if it falls during the octave of Easter, they should not fast. These are the Feast Days which should be Observed in the House of the Temple

75. The nativity of Our Lord; the feast of St Stephen; St John the Evangelist; the Holy Innocents; the eighth day of Christmas, which is New Year's Day; Epiphany; St Mary Candlemas; St Mathias the Apostle; the Annunciation of Our Lady in March; Easter and the three days following; St George; Sts Philip and James, two apostles; the finding of the Holy Cross; the Ascension of Our Lord; Pentecost and the two days following; St John the Baptist; St Peter and St Paul, two apostles: St Mary Magdalene; St James the Apostle; St Laurence; the Assumption of Our Lady; the nativity of Our Lady; the Exaltation of the Holy Cross; St Matthew the Apostle; St Michael; Sts Simon and Jude; the feast of All Saints; St Martin in winter; St Catherine in winter; St Andrew; St Nicholas in winter; St Thomas the Apostle. 76. None of the lesser feasts should be kept by the house of the Temple. And we wish and advise that this be strictly kept and adhered to: that all the brothers of the Temple should fast from the Sunday before St Martin's to the nativity of Qur Lord, unless illness prevents them. And if it happens that the feast of St Martin falls on a Sunday, the brothers should go without meat on the preceding Sunday

Copyright (C) 1992, J. M. Upton-Ward. Excerpted here by kind permission of the author. This file may be copied on the condition that the entire contents, including the header and this copyright notice, remain intact. The Primitive Rule of the Templars Trans. Mrs. Judith Upton-Ward (Reprinted by kind permission of the author)

The Text of the Magna Carta

Chapter 31

I am inclosing the Magna Carta wording from 1215 AD. It is an interesting read, just about the doings of Robert alone. I can only guess that most everybody could not read for it themselves at that time in history, as most were illiterate. So then, most would not know just what was being forced upon King John in the Magna Carta, nor anything about the real players in the First Barons Revolt unless they themselves were involved. It did not take long for the Magna Carta to be amended, mostly for obvious reasons. The Text of this Document, after reading the Interconnections of the Barons of the Magna Carta Chapter above, changes everything about this document, called "The Magna Carta."

Please remember there was at that time no standardization of the English Language. Included in this document are Articles pertaining to the Barons involved that have nothing to do with limiting the power of a King. Showing that these Barons were working at self-service first and service to the public second.

Also, note Thomas Basset, Alan Basset mention in the second paragraph, these 2 men are the predecessors to Elizabeth de Bisset mentioned in the first chapter of, Robert.

Text of the Magna Carta…dated 1215 JOHN, by the grace of God King of England, Lord of Ireland, Duke of Normandy and Aquitaine, and Count of Anjou, to his archbishops, bishops, abbots, earls, barons, justices, foresters, sheriffs, stewards, servants, and to all his officials and loyal subjects, Greeting. KNOW THAT BEFORE GOD, for the health of our soul and those of our ancestors and heirs, to the honour of God, the exaltation of the holy Church, and the better ordering of our kingdom, at the advice of our reverend fathers Stephen, archbishop of Canterbury, primate of all England, and cardinal of the holy Roman Church, Henry archbishop of Dublin, William bishop of London, Peter bishop of Winchester, Jocelin bishop of Bath and Glastonbury, Hugh bishop of Lincoln, Walter Bishop of Worcester, William bishop of Coventry, Benedict bishop of Rochester, Master Pandulf subdeacon and member of the papal household, Brother Aymeric master of the knighthood of the Temple in England, William Marshal earl of Pembroke, William earl of Salisbury, William earl of Warren, William earl of Arundel, Alan de Galloway constable of Scotland, Warin Fitz Gerald, Peter Fitz Herbert, Hubert de Burgh seneschal of Poitou, Hugh de Neville, Matthew Fitz Herbert, Thomas Basset, Alan Basset, Philip Daubeny, Robert de Roppeley, John Marshal, John Fitz Hugh, and other loyal subjects: +

(1) FIRST, THAT WE HAVE GRANTED TO GOD, and by this present charter have confirmed for us and our heirs in perpetuity, that the English Church shall be free, and shall have its rights undiminished, and its liberties unimpaired. That we wish this so to be observed, appears from the fact that of our own free will, before the outbreak of the present dispute between us and our barons, we granted and confirmed by charter the freedom of the Church's elections - a right reckoned to be of the greatest necessity and importance to it - and caused this to be confirmed by Pope Innocent III. This freedom we shall observe ourselves, and desire to be observed in good faith by our heirs in perpetuity. TO ALL FREE MEN OF OUR KINGDOM we have also granted, for us and our heirs for ever, all the liberties written out below, to have and to keep for them and their heirs, of us and our heirs:

(2) If any earl, baron, or other person that holds lands directly of the Crown, for military service, shall die, and at his death his heir shall be of full age and owe a `relief', the heir shall have his inheritance on payment of the ancient scale of `relief'. That is to say, the heir or heirs of an earl shall pay £100 for the entire earl's barony, the heir or heirs of a knight 100s. at most for the entire knight's `fee', and any man that owes less shall pay less, in accordance with the ancient usage of `fees'

(3) But if the heir of such a person is under age and a ward, when he comes of age he shall have his inheritance without 'relief' or fine.

(4) The guardian of the land of an heir who is under age shall take from it only reasonable revenues, customary dues, and feudal services. He shall do this without destruction or damage to men or property. If we have given the guardianship of the land to a sheriff, or to any person answerable to us for the revenues, and he commits destruction or damage, we will exact compensation from him, and the land shall be entrusted to two worthy and prudent men of the same 'fee', who shall be answerable to us for the revenues, or to the person to whom we have assigned them. If we have given or sold to anyone the guardianship of such land, and he causes destruction or damage, he shall lose the guardianship of it, and it shall be handed over to two worthy and prudent men of the same 'fee', who shall be similarly answerable to us.

(5) For so long as a guardian has guardianship of such land, he shall maintain the houses, parks, fish preserves, ponds, mills, and everything else pertaining to it, from the revenues of the land itself. When the heir comes of age, he shall restore the whole land to him, stocked with plough teams and such implements of husbandry as the season demands and the revenues from the land can reasonably bear.

(6) Heirs may be given in marriage, but not to someone of lower social standing. Before a marriage takes place, it shall be' made known to the heir's next-of-kin.***

(7) At her husband's death, a widow may have her marriage portion and inheritance at once and without trouble. She shall pay nothing for her dower, marriage portion, or any inheritance that she and her husband held jointly on the day of his death. She may remain in her husband's house for forty days after his death, and within this period her dower shall be assigned to her.***

(8) No widow shall be compelled to marry, so long as she wishes to remain without a husband. But she must give security that she will not marry without royal consent, if she holds her lands of the Crown, or without the consent of whatever other lord she may hold them of.***

(9) Neither we nor our officials will seize any land or rent in payment of a debt, so long as the debtor has movable goods sufficient to discharge the debt. A debtor's sureties shall not be distrained upon so long as the debtor himself can discharge his debt. If, for lack of means, the debtor is unable to discharge his debt, his sureties shall be answerable for it. If they so desire, they may have the debtor's lands and rents until they have received satisfaction for the debt that they paid for him, unless the debtor can show that he has settled his obligations to them. *

(10) If anyone who has borrowed a sum of money from Jews dies before the debt has been repaid, his heir shall pay no interest on the debt for so long as he remains under age, irrespective of whom he holds his lands. If such a debt falls into the hands of the Crown, it will take nothing except the principal sum specified in the bond. *

(11) If a man dies owing money to Jews, his wife may have her dower and pay nothing towards the debt from it. If he leaves children that are under age, their needs may also be provided for on a scale appropriate to the size of his holding of lands. The debt is to be paid out of the residue, reserving the service due to his feudal lords. Debts owed to persons other than Jews are to be dealt with similarly. *

(12) No `scutage' or `aid' may be levied in our kingdom without its general consent, unless it is for the ransom of our person, to make our eldest son a knight, and (once) to marry our eldest daughter. For these purposes ouly a reasonable `aid' may be levied. `Aids' from the city of London are to be treated similarly. +

(13) The city of London shall enjoy all its ancient liberties and free customs, both by land and by water. We also will and grant that all other cities, boroughs, towns, and ports shall enjoy all their liberties and free customs. *

(14) To obtain the general consent of the realm for the assessment of an `aid' - except in the three cases specified above - or a `scutage', we will cause the archbishops, bishops, abbots, earls, and greater barons to be summoned individually by letter. To those who hold lands directly

of us we will cause a general summons to be issued, through the sheriffs and other officials, to come together on a fixed day (of which at least forty days notice shall be given) and at a fixed place. In all letters of summons, the cause of the summons will be stated. When a summons has been issued, the business appointed for the day shall go forward in accordance with the resolution of those present, even if not all those who were summoned have appeared. *

(15) In future we will allow no one to levy an `aid' from his free men, except to ransom his person, to make his eldest son a knight, and (once) to marry his eldest daughter. For these purposes only a reasonable `aid' may be levied.

(16) No man shall be forced to perform more service for a knight's `fee', or other free holding of land, than is due from it.

(17) Ordinary lawsuits shall not follow the royal court around, but shall be held in a fixed place.

(18) Inquests of novel disseisin, mort d'ancestor, and darrein presentment shall be taken only in their proper county court. We ourselves, or in our absence abroad our chief justice, will send two justices to each county four times a year, and these justices, with four knights of the county elected by the county itself, shall hold the assizes in the county court, on the day and in the place where the court meets.

(19) If any assizes cannot be taken on the day of the county court, as many knights and freeholders shall afterwards remain behind, of those who have attended the court, as will suffice for the administration of justice, having regard to the volume of business to be done.

(20) For a trivial offence, a free man shall be fined only in proportion to the degree of his offence, and for a serious offence correspondingly, but not so heavily as to deprive him of his livelihood. In the same way, a merchant shall be spared his merchandise, and a husbandman the implements of his husbandry, if they fall upon the mercy of a royal court. None of these fines shall be imposed except by the assessment on oath of reputable men of the neighbourhood.

(21) Earls and barons shall be fined only by their equals, and in proportion to the gravity of their offence.

(22) A fine imposed upon the lay property of a clerk in holy orders shall be assessed upon the same principles, without reference to the value of his ecclesiastical benefice.

(23) No town or person shall be forced to build bridges over rivers except those with an ancient obligation to do so.

(24) No sheriff, constable, coroners, or other royal officials are to hold lawsuits that should be held by the royal justices. *

(25) Every county, hundred, wapentake, and tithing shall remain at its ancient rent, without increase, except the royal demesne manors.

(26) If at the death of a man who holds a lay `fee' of the Crown, a sheriff or royal official produces royal letters patent of summons for a debt due to the Crown, it shall be lawful for them to seize and list movable goods found in the lay `fee' of the dead man to the value of the debt, as assessed by worthy men. Nothing shall be removed until the whole debt is paid, when the residue shall be given over to the executors to carry out the dead man s will. If no debt is due to the Crown, all the movable goods shall be regarded as the property of the dead man, except the reasonable shares of his wife and children. *

(27) If a free man dies intestate, his movable goods are to be distributed by his next-of-kin and friends, under the supervision of the Church. The rights of his debtors are to be preserved.

(28) No constable or other royal official shall take corn or other movable goods from any man without immediate payment, unless the seller voluntarily offers postponement of this.

(29) No constable may compel a knight to pay money for castle-guard if the knight is willing to undertake the guard in person, or with reasonable excuse to supply some other fit man to do it. A knight taken or sent on military service shall be excused from castle-guard for the period of this servlce.

(30) No sheriff, royal official, or other person shall take horses or carts for transport from any free man, without his consent.

(31) Neither we nor any royal official will take wood for our castle, or for any other purpose, without the consent of the owner.

(32) We will not keep the lands of people convicted of felony in our hand for longer than a year and a day, after which they shall be returned to the lords of the `fees' concerned.

(33) All fish-weirs shall be removed from the Thames, the Medway, and throughout the whole of England, except on the sea coast.

(34) The writ called precipe shall not in future be issued to anyone in respect of any holding of land, if a free man could thereby be deprived of the right of trial in his own lord's court.

(35) There shall be standard measures of wine, ale, and corn (the London quarter), throughout the kingdom. There shall also be a standard width of dyed cloth, russett, and haberject, namely two ells within the selvedges. Weights are to be standardised similarly.

(36) In future nothing shall be paid or accepted for the issue of a writ of inquisition of life or limbs. It shall be given gratis, and not refused.

(37) If a man holds land of the Crown by `fee-farm', `socage', or `burgage', and also holds land of someone else for knight's service, we will not have guardianship of his heir, nor of the land that belongs to the other person's `fee', by virtue of the `fee-farm', `socage', or `burgage', unless the `fee-farm' owes knight's service. We will not have the guardianship of a man's heir, or of land that he holds of someone else, by reason of any small property that he may hold of the Crown for a service of knives, arrows, or the like.

(38) In future no official shall place a man on trial upon his own unsupported statement, without producing credible witnesses to the truth of it. +

(39) No free man shall be seized or imprisoned, or stripped of his rights or possessions, or outlawed or exiled, or deprived of his standing in any other way, nor will we proceed with force against him, or send

others to do so, except by the lawful judgement of his equals or by the law of the land. +

(40) To no one will we sell, to no one deny or delay right or justice.

(41) All merchants may enter or leave England unharmed and without fear, and may stay or travel within it, by land or water, for purposes of trade, free from all illegal exactions, in accordance with ancient and lawful customs. This, however, does not apply in time of war to merchants from a country that is at war with us. Any such merchants found in our country at the outbreak of war shall be detained without injury to their persons or property, until we or our chief justice have discovered how our own merchants are being treated in the country at war with us. If our own merchants are safe they shall be safe too. *

(42) In future it shall be lawful for any man to leave and return to our kingdom unharmed and without fear, by land or water, preserving his allegiance to us, except in time of war, for some short period, for the common benefit of the realm. People that have been imprisoned or outlawed in accordance with the law of the land, people from a country that is at war with us, and merchants - who shall be dealt with as stated above - are excepted from this provision.

(43) If a man holds lands of any 'escheat' such as the 'honour' of Wallingford, Nottingham, Boulogne, Lancaster, or of other 'escheats' in our hand that are baronies, at his death his heir shall give us only the 'relief' and service that he would have made to the baron, had the barony been in the baron's hand. We will hold the 'escheat' in the same manner as the baron held it.

(44) People who live outside the forest need not in future appear before the royal justices of the forest in answer to general summonses, unless they are actually involved in proceedings or are sureties for someone who has been seized for a forest offence. *

(45) We will appoint as justices, constables, sheriffs, or other officials, only men that know the law of the realm and are minded to keep it well.

(46) All barons who have founded abbeys, and have charters of English kings or ancient tenure as evidence of this, may have guardianship of them when there is no abbot, as is their due.

(47) All forests that have been created in our reign shall at once be disafforested. River-banks that have been enclosed in our reign shall be treated similarly. *

(48) All evil customs relating to forests and warrens, foresters, warreners, sheriffs and their servants, or river-banks and their wardens, are at once to be investigated in every county by twelve sworn knights of the county, and within forty days of their enquiry the evil customs are to be abolished completely and irrevocably. But we, or our chief justice if we are not in England, are first to be informed. *

(49) We will at once return all hostages and charters delivered up to us by Englishmen as security for peace or for loyal service. *

(50) We will remove completely from their offices the kinsmen of Gerard de Athée, and in future they shall hold no offices in England. The people in question are Engelard de Cigogné', Peter, Guy, and Andrew de Chanceaux, Guy de Cigogné, Geoffrey de Martigny and his brothers, Philip Marc and his brothers, with Geoffrey his nephew, and all their followers. *

(51) As soon as peace is restored, we will remove from the kingdom all the foreign knights, bowmen, their attendants, and the mercenaries that have come to it, to its harm, with horses and arms. *

(52) To any man whom we have deprived or dispossessed of lands, castles, liberties, or rights, without the lawful judgement of his equals, we will at once restore these. In cases of dispute the matter shall be resolved by the judgement of the twenty-five barons referred to below in the clause for securing the peace (§ 61). In cases, however, where a man was deprived or dispossessed of something without the lawful judgement of his equals by our father King Henry or our brother King Richard, and it remains in our hands or is held by others under our warranty, we shall have respite for the period commonly allowed to Crusaders, unless a lawsuit had been begun, or an enquiry had been made at our order, before we took the Cross as a Crusader. On our

return from the Crusade, or if we abandon it, we will at once render justice in full. *

(53) We shall have similar respite in rendering justice in connexion with forests that are to be disafforested, or to remain forests, when these were first aforested by our father Henry or our brother Richard; with the guardianship of lands in another person's 'fee', when we have hitherto had this by virtue of a 'fee' held of us for knight's service by a third party; and with abbeys founded in another person's 'fee', in which the lord of the 'fee' claims to own a right. On our return from the Crusade, or if we abandon it, we will at once do full justice to complaints about these matters.

(54) No one shall be arrested or imprisoned on the appeal of a woman for the death of any person except her husband. *

(55) All fines that have been given to us unjustiy and against the law of the land, and all fines that we have exacted unjustly, shall be entirely remitted or the matter decided by a majority judgement of the twenty-five barons referred to below in the clause for securing the peace (§ 61) together with Stephen, archbishop of Canterbury, if he can be present, and such others as he wishes to bring with him. If the archbishop cannot be present, proceedings shall continue without him, provided that if any of the twenty-five barons has been involved in a similar suit himself, his judgement shall be set aside, and someone else chosen and sworn in his place, as a substitute for the single occasion, by the rest of the twenty-five.

(56) If we have deprived or dispossessed any Welshmen of lands, liberties, or anything else in England or in Wales, without the lawful judgement of their equals, these are at once to be returned to them. A dispute on this point shall be determined in the Marches by the judgement of equals. English law shall apply to holdings of land in England, Welsh law to those in Wales, and the law of the Marches to those in the Marches. The Welsh shall treat us and ours in the same way. *

(57) In cases where a Welshman was deprived or dispossessed of anything, without the lawful judgement of his equals, by our father King Henry or our brother King Richard, and it remains in our hands or is held by others under our warranty, we shall have respite for the period commonly allowed to Crusaders, unless a lawsuit had been begun, or an enquiry had been made at our order, before we took the Cross as a Crusader. But on our return from the Crusade, or if we abandon it, we will at once do full justice according to the laws of Wales and the said regions. *

(58) We will at once return the son of Llywelyn, all Welsh hostages, and the charters delivered to us as security for the peace. *

(59) With regard to the return of the sisters and hostages of Alexander, king of Scotland, his liberties and his rights, we will treat him in the same way as our other barons of England, unless it appears from the charters that we hold from his father William, formerly king of Scotland, that he should be treated otherwise. This matter shall be resolved by the judgement of his equals in our court.

(60) All these customs and liberties that we have granted shall be observed in our kingdom in so far as concerns our own relations with our subjects. Let all men of our kingdom, whether clergy or laymen, observe them similarly in their relations with their own men. *

(61) SINCE WE HAVE GRANTED ALL THESE THINGS for God, for the better ordering of our kingdom, and to allay the discord that has arisen between us and our barons, and since we desire that they shall be enjoyed in their entirety, with lasting strength, for ever, we give and grant to the barons the following security: The barons shall elect twenty-five of their number to keep, and cause to be observed with all their might, the peace and liberties granted and confirmed to them by this charter. If we, our chief justice, our officials, or any of our servants offend in any respect against any man, or transgress any of the articles of the peace or of this security, and the offence is made known to four of the said twenty-five barons, they shall come to us - or in our absence from the kingdom to the chief justice - to declare it and claim immediate redress. If we, or in our absence abroad the chief

justice, make no redress within forty days, reckoning from the day on which the offence was declared to us or to him, the four barons shall refer the matter to the rest of the twenty-five barons, who may distrain upon and assail us in every way possible, with the support of the whole community of the land, by seizing our castles, lands, possessions, or anything else saving only our own person and those of the queen and our children, until they have secured such redress as they have determined upon. Having secured the redress, they may then resume their normal obedience to us. Any man who so desires may take an oath to obey the commands of the twenty-five barons for the achievement of these ends, and to join with them in assailing us to the utmost of his power. We give public and free permission to take this oath to any man who so desires, and at no time will we prohibit any man from taking it. Indeed, we will compel any of our subjects who are unwilling to take it to swear it at our command. If one of the twenty-five barons dies or leaves the country, or is prevented in any other way from discharging his duties, the rest of them shall choose another baron in his place, at their discretion, who shall be duly sworn in as they were. *** In the event of disagreement among the twenty-five barons on any matter referred to them for decision, the verdict of the majority present shall have the same validity as a unanimous verdict of the whole twenty-five, whether these were all present or some of those summoned were unwilling or unable to appear. The twenty-five barons shall swear to obey all the above articles faithfully, and shall cause them to be obeyed by others to the best of their power. We will not seek to procure from anyone, either by our own efforts or those of a third party, anything by which any part of these concessions or liberties might be revoked or diminished. Should such a thing be procured, it shall be null and void and we will at no time make use of it, either ourselves or through a third party. *

(62) We have remitted and pardoned fully to all men any ill-will, hurt, or grudges that have arisen between us and our subjects, whether clergy or laymen, since the beginning of the dispute. We have in addition remitted fully, and for our own part have also pardoned, to all clergy and laymen any offences committed as a result of the said dispute

between Easter in the sixteenth year of our reign (i.e. 1215) and the restoration of peace. In addition we have caused letters patent to be made for the barons, bearing witness to this security and to the concessions set out above, over the seals of Stephen archbishop of Canterbury, Henry archbishop of Dublin, the other bishops named above, and Master Pandulf. *

(63) IT IS ACCORDINGLY OUR WISH AND COMMAND that the English Church shall be free, and that men in our kingdom shall have and keep all these liberties, rights, and concessions, well and peaceably in their fullness and entirety for them and their heirs, of us and our heirs, in all things and all places for ever. Both we and the barons have sworn that all this shall be observed in good faith and without deceit. Witness the abovementioned people and many others. Given by our hand in the meadow that is called Runnymede, between Windsor and Staines, on the fifteenth day of June in the seventeenth year of our reign 1215.

Bibliography

Charles Addison, The History of The Knights Templar, The Temple Church, and The Temple, London: Longman, Brown, Green, and Longman Paternoster Row. 1842

Advocate.com , Newsmagazine Pride Media, United States 1967-2020

George Bain, History of Nairnshire, Nairn Scotland Telegraph Office Nairn 1893

https://www.britannica.com/event/Siege-of-Jerusalem-70 Hugh Chisholm, Encyclopædia Britannica

"Rose, George" Scotland 1911

Arthur Garfunkel, Paul Simon, Scarborough Fair/Canticle lyrics, Universal Music Publishing Group

Judith A. Green, The Aristocracy of Norman England, Cambridge: The Press syndicate of the University of Cambridge 1997

Richard Gough, Sepulchral Monuments In Britain, London: J. Nichols 1788

Thomas Duffas Hardy, Rotuli Chartarum in Turri londinensi Asservati AB Anno 1199- 1216 Volume 1,

 1837 C E Kempe & Co. depicting Saint Michael Archangel slaying the devil. Alamy Stock Photo License details Country Worldwide Usage Magazines and books

John Kenrick, A Selection of Papers on Subjects of Archaeology and History Communicated to the Yorkshire Philosophical Society, London: Longman, Green, Longman,Roberts, & Green 1864

George Lawton, The Religious Houses of Yorkshire, London: Simpkin & Co., Stationers' Hall Court 1853

The National Archives (.gov) uk 1215 The Magna Carta

The Norman People and Their Existing Descendants in The British Dominions and The United States of America, London 1874

Magna Carta, Rummymeade: 1215

Mark Anthony Lower, Patronymica Britannica. A Dictionary of the Family Names of the United Kingdom, London John Russell Smith 1860

P. Kyle McCarter, The mystery of the Copper Scroll, in Hershel Shanks, ed., Understanding the Dead Sea Scrolls New York: Random House, 1992

The Manuscripts of His Grace The Duke of Rutland, K.G. Presented to Parliament by Command of His Majesty, London: Mackie & Co. L.D. 59 Fleet Street., E.C. 1905

Martyrdom of St George, Collegiate Church of St. George in Tubingen, Germany on October 21, 2014 Model release NO Property release NO Credit line Zvonimir Atletić / Alamy Stock Photo License details Image use Magazines and books Start date 17 April 2020 Expiry date 17 April 2025 Country Worldwide Additional details Use in a magazine or book (print and/or digital), inside use, 2,500 circulation, worldwide for 5 years (excludes advertising).

https://www.medicalnewstoday.com/ Healthline Media UK LTD. 2004-2020

Douglas Richardson, Kimball G. Everingham, Magna Carta Ancestry, Salt Lake City, Utah USA, Genealogical Publishing Company 2005

George Rose, Diaries and Correspondence of the Right Hon. George Rose, London: Richard Bentley, New Burlington Street, Publisher in Ordinary to Her Majesty, 1860

Hershel Shanks. ed, The mystery of the Copper Scroll, Understanding the Dead Sea Scrolls (New York: Random House, 1992), p. 237.

Walter Scott, Ivanhoe The Project Gutenberg EBook of Ivanhoe, This eBook is for the use of anyone anywhere at no cost and with almost no restrictions whatsoever. You may copy it, give it away or re-use it under the terms of the Project Gutenberg License included with this eBook or online at www.gutenberg.org 1819

Hershel Shanks, The Copper Scroll and the Search for the Temple Treasure, Biblical Archaeology Society 2007

Lachlan Shaw, A Genealogical Deduction of the Family of Rose Kilravock, Nairnshire Scotland: Nairn 1848

Harry Speight, Nidderdale and the Garden of the Nidd: a Yorkshire Rhineland, London: Elliot Stock, 62, Paternoster Row, EC 1894

Thomas Stapleton, Magni Rotuli Scaccarii Normanniae Sub Regibus Angliae, Volume 1, London: J.B. Nicholas and Son 25 Parliament Street 1844

Thomas Stapleton, Magni Rotuli Scaccarii Normanniae Sub Regibus Angliae, Volume 2, London: J.B. Nicholas and Son 25 Parliament Street 1844

University of Mull, History Dept. Edinburgh, School of History. Centre for Medieval and Renaissance Studies. Rules of the Knights Templar Copyright (C) 1992,

J. M. Upton-Ward. Excerpted here by kind permission of the author. This file may be copied on the condition that the entire contents, including the header and this copyright notice, remain intact.

Merriam-Webster, Dictionary, Merriam-Webster, 2020

Roger Wendover, Flowers of History, London: Henry G Bohn, 1849

William Wheater, Some Historic Mansions of Yorkshire and Their Associations, Volume 1, Leeds: Richard Jackson 1888

William Wheater, Some Historic Mansions of Yorkshire and Their Associations, Volume 2, Leeds:

Richard Jackson 1889 University of Mull, History Dept. Edinburgh, School of History. Centre for Medieval and Renaissance Studies.

Made in the USA
San Bernardino, CA
03 June 2020